math expressions

Dr. Karen C. Fuson

Center Planning Guide

Grade 4

This material is based upon work supported by the
National Science Foundation
under Grant Numbers
ESI-9816320, REC-9806020, and RED-935373.

Any opinions, findings, and conclusions, or recommendations expressed in this material are those of the author and do not necessarily reflect the views of the National Science Foundation.

HMH

Printed in the U.S.A.

ISBN 978-1-328-89835-7

3 4 5 6 7 8 9 10 2266 26 25 24 23 22 21 20 19 18

4500710507 B C D E F G

Math Activity Center at a Glance

What are centers?

Centers are resource hubs for a variety of learning activities. They provide an opportunity for students to practice and apply concepts and skills they have acquired in core content or to extend knowledge through exploration. At each center, students work individually or in small groups on a given task for an allotted amount of time. Each center has the materials the student needs to complete the activity, such as connecting cubes, or digital tools, such as an interactive game. Often centers are set up in a rotation, so that students have the opportunity to work on a variety of tasks.

What is the purpose of the Math Expressions Math Activity Center?

The Math Activity Center provides teachers with the materials they need to differentiate instruction, and it provides students with a variety of differentiated learning experiences. The number of activities provided is more than any one student could use in a school year. By choosing the activities that best fit the needs of their students, teachers can use the Math Activity Center to create a customized learning environment within their classrooms.

Three Centers in One

The Math Activity Center Tri-Fold displays the print components in an organized central location. Designate an area of the classroom to display the tri-fold so that it is accessible to all students. Tri-fold contents are:

- Games: game boards for practice; one per unit
- Math Readers: fiction and non-fiction stories for math connections; one per unit
- Math Activity Cards and Writing Prompts: circle cards for intervention, triangle cards for on-level, and square cards for challenge activities; one each per lesson
- In-Depth Inquiry Based Tasks cards: individual and group cumulative tasks; one per unit
- Practice Reteach Challenge: reproducible activity sheets; on page of each type per lesson
- Center Planning Guide: information and lesson plans assistance for the teacher

It may be helpful to place current unit materials in the tri-fold and store additional materials.

Using the Online Math Activity Center teacher dashboard allows you to view content and assign activities to individual students and provides easy access for reproducible materials also available with the tri-fold. Teachers can search for all assets related to a particular unit and/or lesson, or they can access full components to browse content.

- Printable content available only to the teacher: Practice, Reteach, Challenge, RtI Tier 3, Game boards, Math Activity Cards, Center Planning Guide
- Components that are available to the student through assignment only: RtI Tier 1 and Tier 2
- Free-play components that mimic student experience: Online Fluency Builders, interactive Games, PMT Lesson Checks, Math Readers

The Online Math Activity Center student dashboard allows students to engage freely in Online Fluency Builders, PMT Lesson Checks, interactive Games, and Math Readers.

How do I group centers?

Centers allow for many different configurations of student learning environments. Students can work independently, in groups they choose based on interest, and in mixed ability or leveled groups. Each center may be configured differently as you design it to be. Classroom size, student interest, and grade level will help determine how to customize your centers. Centers allow you to differentiate learning to the needs of your students. Use the Math Activity Center Assignments BLM to organize and communicate center assignments.

Center Tips:

- Communicate center procedures and expectations until they have become routine.
- Group students based on the nature and purpose of each center's activity.
- Ensure students know the purpose of each center. Post directions as needed.
- Clearly state directions on where to go after completing the current center.
- Allow time for students to return center materials to their original place.

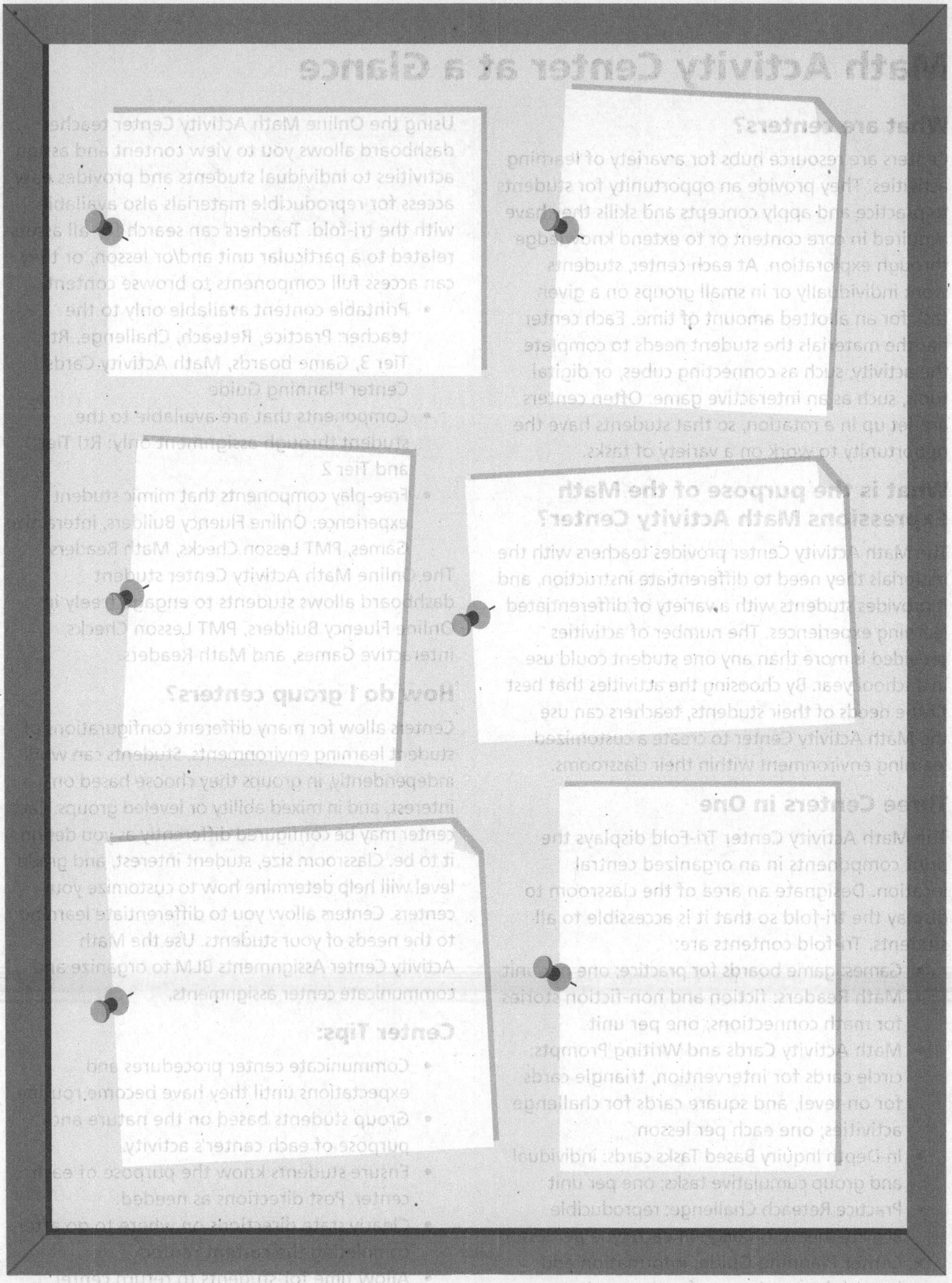

Math Activity Center Components

Math Activity Center Tri-fold

The Math Activity Center Tri-Fold displays the print components in an organized central location. Designate an area of the classroom to display the tri-fold so that it is accessible to all students. Tri-fold contents include unit board games, unit math readers, lesson specific Math Activity Cards and Writing Prompts, unit specific In-Depth Inquiry Based Task cards, Practice Reteach Challenge activity sheets, and the Center Planning Guide.

Online Math Activity Center

- Printable content available only to the teacher: Practice, Reteach, Challenge, RtI Tier 3, Game boards, Math Activity Cards, Center Planning Guide.
- Components that are available to the student through assignment only: RtI Tier 1 and Tier 2
- Free-play components that mimic student experience: Online Fluency Builders, interactive Games, PMT Lesson Checks, Math Readers

Practice, Reteach, Challenge

One task is provided per lesson at each level of differentiation. Interactive student versions are free-play for Practice and Challenge and assignable for all three activity types. Reproducible pages are available online.

Response to Intervention

RtI is provided for each Big Idea. Assignable interactive activities are provided for students requiring Tier 1 and Tier 2 intervention. Diagnostic Assessment and teacher support for Tier 3 intervention are provided as reproducible pages online.

Online Fluency Builders

Online fluency practice is available to support fluency lessons and standards as free-play activities for students.

Games

Interactive Games Interactive student versions of Poggles MX are available as free-play activities, and individual game levels are assignable for correlating lessons.

Board Games Board games are also provided for each unit and reproducible versions of the game boards are available online. Associated copy masters for games requiring cards and spinner are provided in the Center Planning Guide.

PMT Lesson Checks

Personal Math Trainer® Powered by Knewton™ (PMT) intervention/enrichment system components are provided for each lesson. PMT provides practice opportunities for every lesson as well as review.

Math Activity Center Cards

Activity Cards One task is provided per lesson for each level of differentiation. Laminated cards are available for each task and reproducible versions of the activities are available online.

In-Depth Inquiry Based Tasks One task is provided per unit. Instruction for each student or small groups is available on the laminated cards and online. Lesson plans and associated copy masters are provided in the Center Planning Guide

Math Readers

One Math Reader is provided per unit. Picture books are provided and audio-enhanced eBooks are available online.

Math Activity Center Planning Guide

The Center Planning Guide provides instruction to the teacher on how to manage Math Activity Center components for the best possible differentiation within the classroom. A printable version is available online.

Using the Math Activity Center

Content from the Math Activity Center can be accessed online through a link in the teacher dashboard. Teachers have access to teacher and student views of the content. Math Activity Center content is also available in a handy kit for the classroom.

Teacher View

Viewing the Content

Program components are organized to allow for teachers to access the content in two different ways. The teacher who wants to see all of the differentiated instruction support for an individual lesson can use the Unit and Lesson filters to display that content. The teacher who prefers to look at a particular component in more depth can select the component type from the Teacher View landing page.

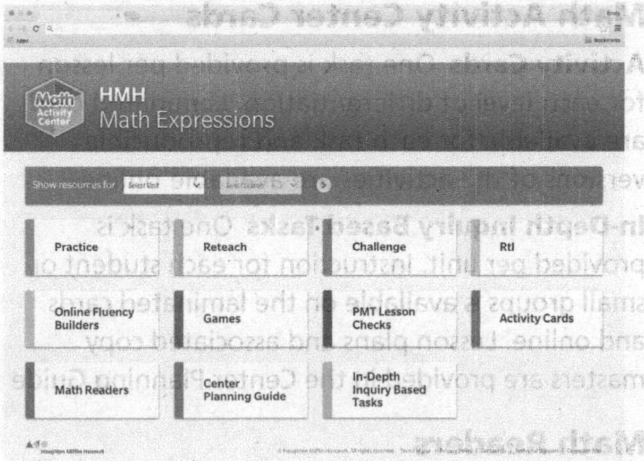

Viewing by Lesson

Teachers can scroll through all available resources or filter to specific resources at the lesson level. They can also easily navigate forward and backward to additional lessons.

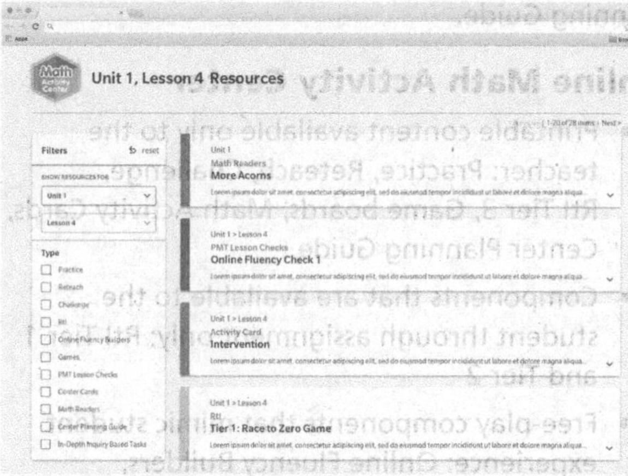

Viewing by Component

Selecting a component from the Teacher View landing page allows teachers to filter within individual components.

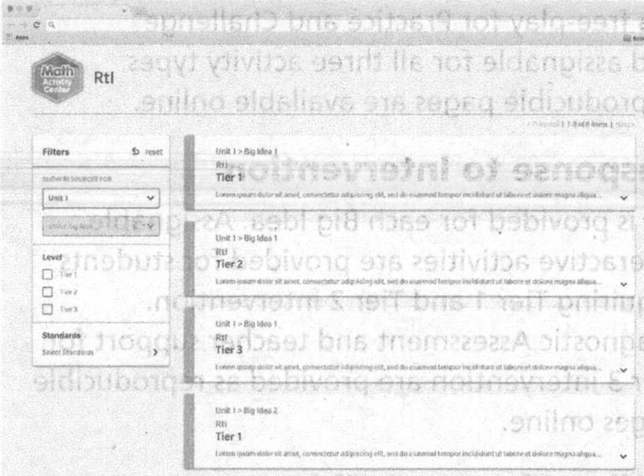

Preview and Select

Teachers can preview and select individual activities or pages for planning purposes or classroom use.

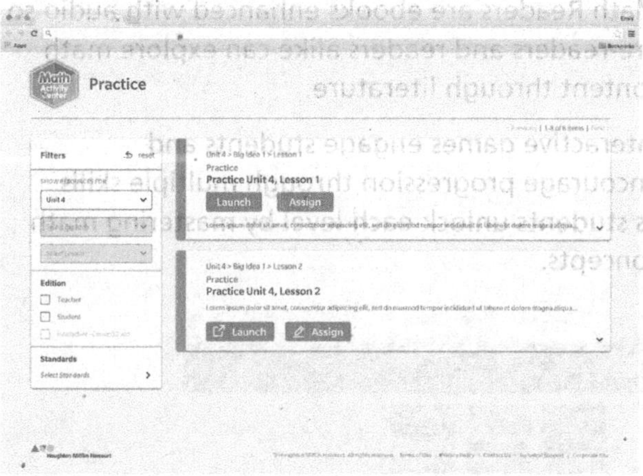

Assigning

Teachers can easily assign activities from the Math Activity Center. By assigning activities, you customize learning for each student. Students only have access to specific items after the teacher has assigned it, such as Reteach, RtI Tier 1, and RtI Tier 2. Other Math Activity Center components available only to the teacher have PDF versions that can printed and given to students.

Student View

Viewing the Content

Students have access to a number of free-play activities within the Math Activity Center. Intervention and Reteach materials are not made available via free-play, as it is expected that the teacher will assign those materials as needed. The student view has a design that allows even the youngest students access to and use of differentiated instruction supplemental materials easily.

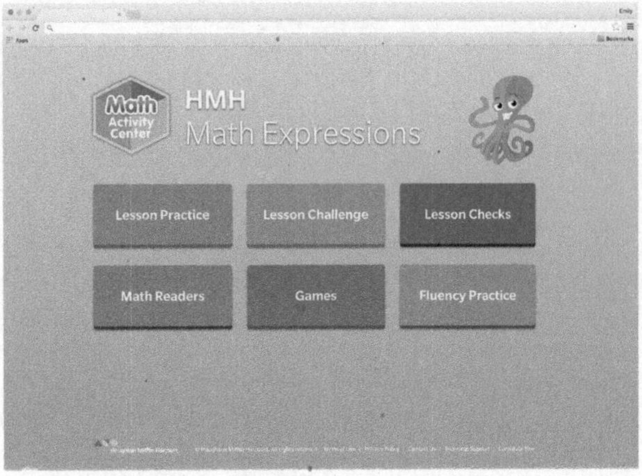

Games and Math Readers

Access to interactive Games and Math Readers are a touch away from the Student View landing page.

Math Readers are ebooks enhanced with audio so pre-readers and readers alike can explore math content through literature.

Interactive games engage students and encourage progression through multiple skills as students unlock each level by mastering math concepts.

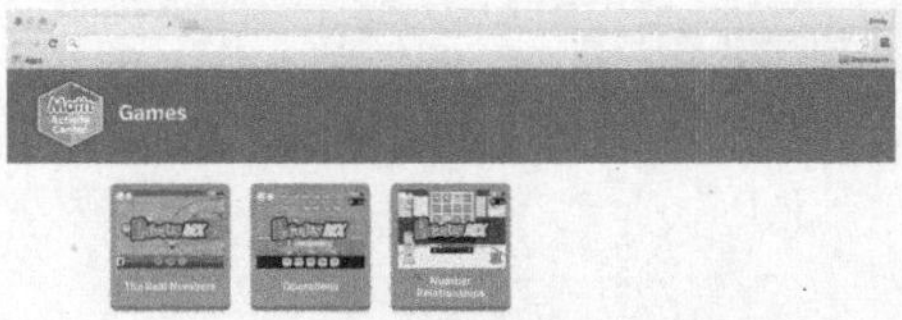

Additional Lesson Support

Fluency Practice, Lesson Practice, Lesson Challenge, and Lesson Check activities are also provided to supplement the program.

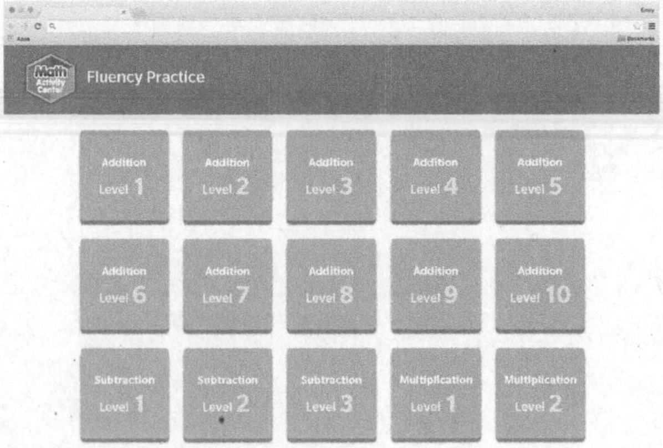

Using the Tri-Fold

The physical Math Activity Center is a convenient tri-fold display for use within the classroom.

Classroom Set-up

Remove the Center Planning Guide before displaying the tri-fold for students. You may choose to keep all other components in the tri-fold throughout the year or display only the pages and components relevant to your current instruction.

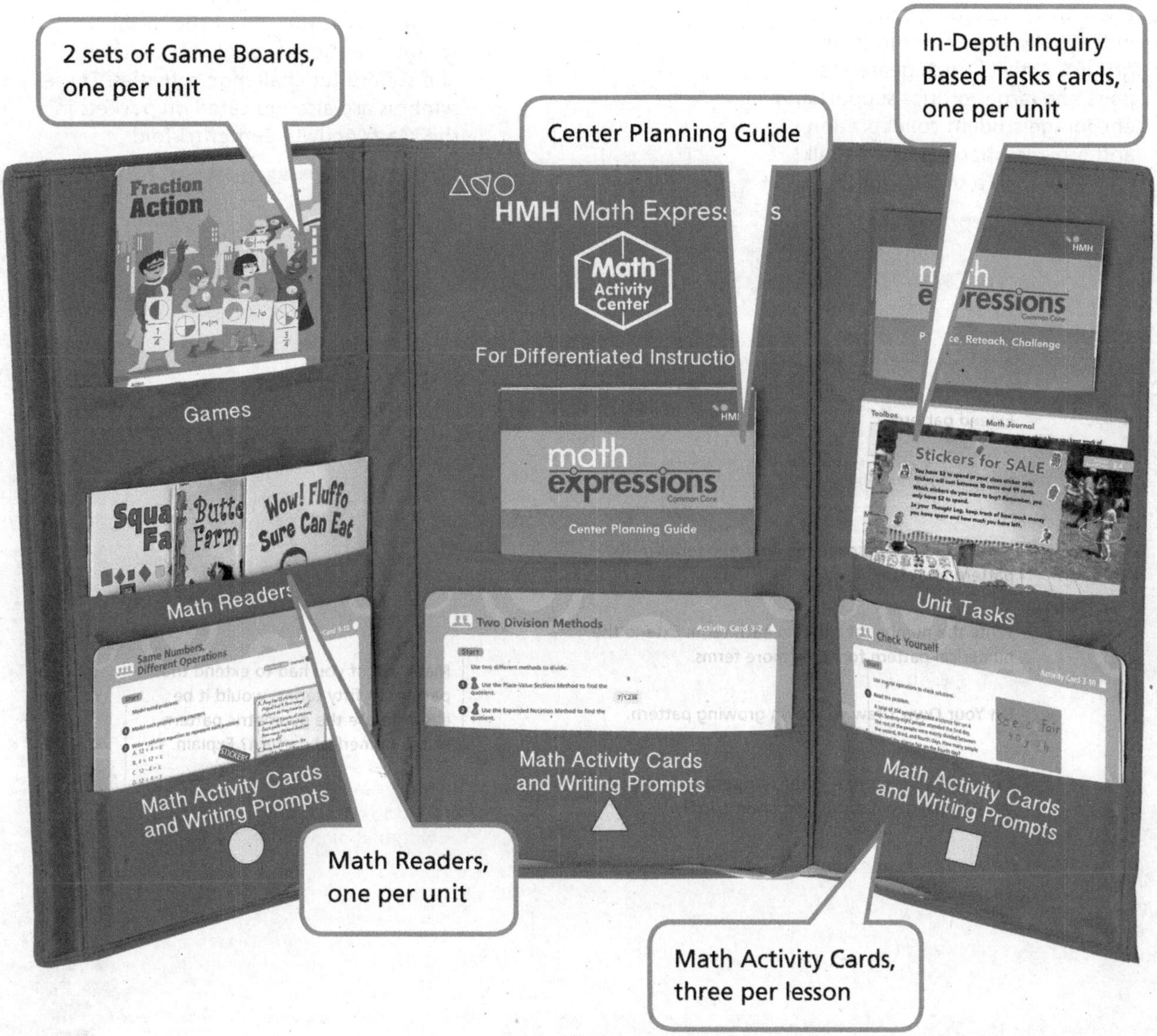

Using the Activity Cards

The Activity Cards in the Math Activity Center help to reinforce and extend mathematical concepts and skills taught in the Math Expressions lessons. In the Math Expressions Teacher Edition you will see references to the Activity Cards in the Differentiated Instruction section of the lesson.

The Activity Cards indicate whether the activity should be completed individually, in pairs, or in small groups. Activity cards geared to pairs and larger groups support and encourage student collaboration and provide opportunities to talk about math in a smaller setting.

The Activity Cards include intervention, on-level, and challenge activities indicated by symbols at the top of the cards: circle for intervention, triangle for on-level, and square for challenge activities. These symbols are also indicated on pockets in the Math Activity Center tri-fold.

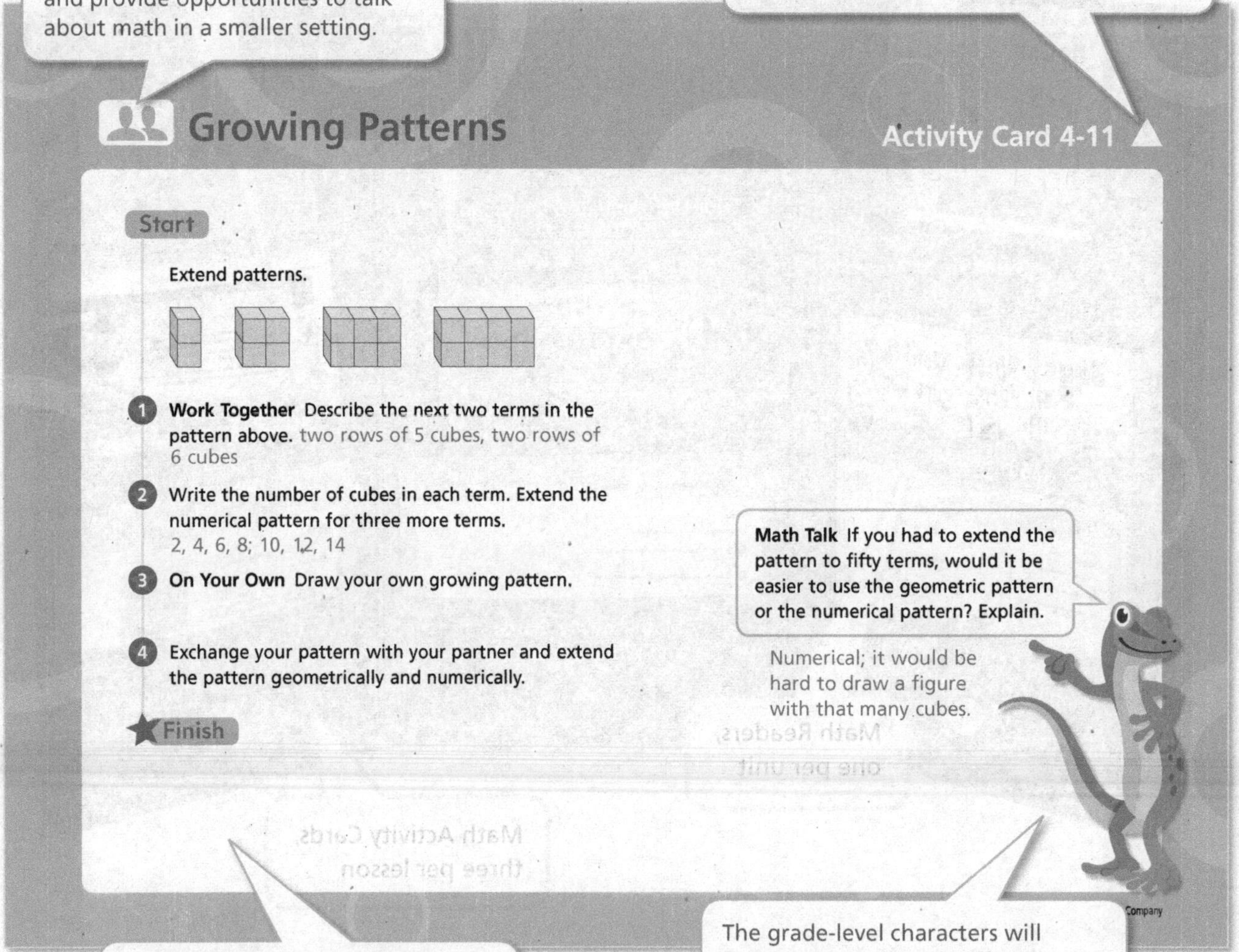
Growing Patterns

Activity Card 4-11 ▲

Start

Extend patterns.

1 **Work Together** Describe the next two terms in the pattern above. two rows of 5 cubes, two rows of 6 cubes

2 Write the number of cubes in each term. Extend the numerical pattern for three more terms. 2, 4, 6, 8; 10, 12, 14

3 **On Your Own** Draw your own growing pattern.

4 Exchange your pattern with your partner and extend the pattern geometrically and numerically.

Finish

Math Talk If you had to extend the pattern to fifty terms, would it be easier to use the geometric pattern or the numerical pattern? Explain.

Numerical; it would be hard to draw a figure with that many cubes.

Activities are clearly stepped out allowing students to work independently from the teacher. Many activities have illustrations that serve as examples of what students should do when working through the activity card.

The grade-level characters will appear on many cards offering hints, reminders, or additional information that will help in successfully completing the math activity.

Using the Activity Cards *(continued)*

- Depending on how often you will be using the Math Activity Center in your classroom, you may want to insert the cards in the Math Activity Center that correspond to the daily lesson or the week of lessons.
- Activities encourage students to write and talk about math, explain their reasoning, and describe more than one solution to problems.
- Access to Teacher Resource Book (TRB) pages is required in some Activity Cards. Prepare a location where students can easily retrieve the needed pages.
- Most of the materials needed for the activities are readily available in the classroom, for example, paper clips, counters, and rulers. You may wish to add these materials to the location where the TRB pages are.

On the back of every Activity Card appears a Writing Prompt, which is at the same level as the activity on the front. The Writing Prompt encourages students to write about math and provides the opportunity to explain their thinking by drawing, writing or both. This opportunity for in-depth thinking and analysis helps prepare students for important assessments.

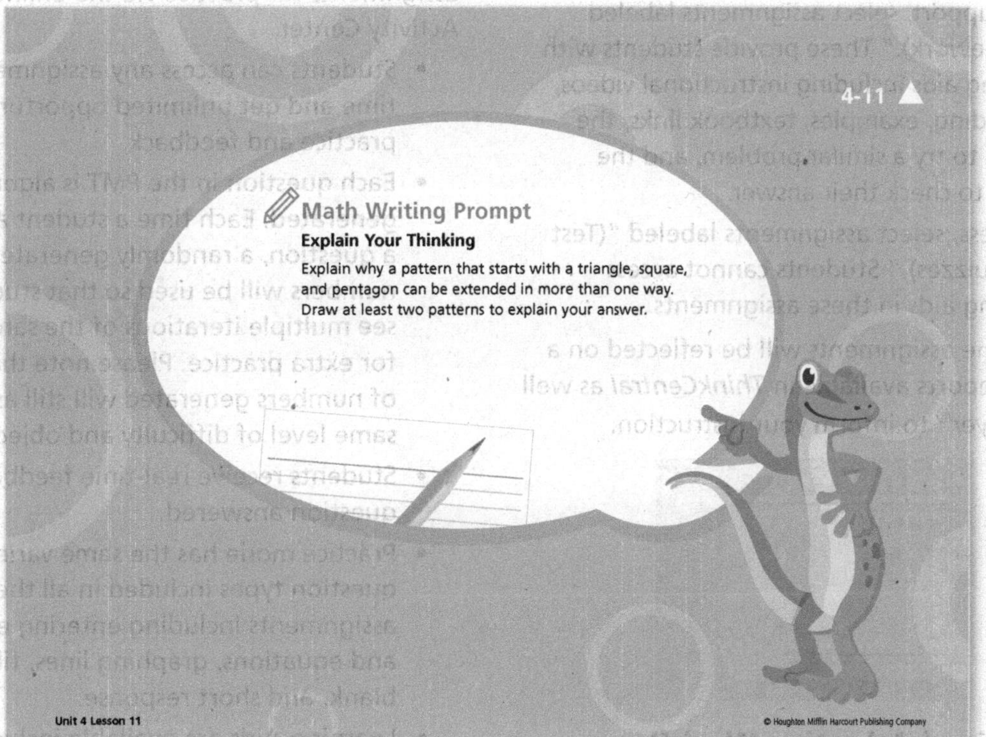

Using the Personal Math Trainer

Tracking Student Progress

The Personal Math Trainer® (PMT) has hundreds of digital assignments aligned to Math Expressions lessons that teachers can administer to students. The data from these assignments provide immediate visibility into what each student knows and is ready to learn, enabling teachers to adjust and differentiate their instruction.

To assign a PMT assignment, log into the *ThinkCentral* dashboard and select the "PMT and Fluency Assignment" icon in the Resources section. Select "Assign" next to the content you want to assign. You can also assign PMT assignments via *HMH Player*®.

- To provide opportunities for practice along with support, select assignments labeled "(Homework)." These provide students with learning aids including instructional videos, scaffolding, examples, textbook links, the option to try a similar problem, and the ability to check their answer.
- To assess, select assignments labeled "(Test and Quizzes)." Students cannot access learning aids in these assignments.

Results of the assignments will be reflected on a variety of reports available in *ThinkCentral* as well as *HMH Player*® to inform your instruction.

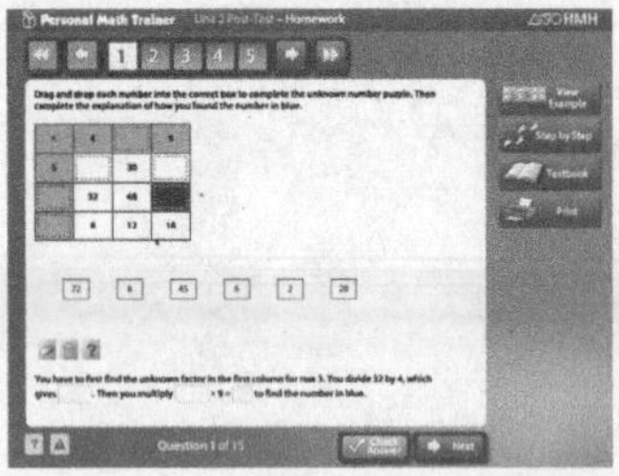

Personalized Adaptive Workflows

The Personal Math Trainer® partners with Knewton™, a leader in adaptive learning technology, to provide a personalized learning experience for each student. Two types of adaptive workflows can be applied to any PMT assignment.

Daily Intervention and Enrichment

- **Step 1:** Up to 5 minutes of personalized adaptive warmup on prerequisite concepts for students who need it.
- **Step 2:** Main assignment.
- **Step 3:** Up to 10 minutes of personalized adaptive enrichment on upcoming skills for students who master the main goals.

Personal Study Plan

- **Step 1:** Main assignment.
- **Step 2:** Up to 15 minutes of personalized adaptive intervention for students who do not master the main goals.
- **Step 3:** Students who received an intervention re-take the main assignment.

Unlimited Practice

Students have unlimited access to PMT assignments for practice via the online Math Activity Center.

- Students can access any assignment at any time and get unlimited opportunities for practice and feedback.
- Each question in the PMT is algorithmically generated. Each time a student accesses a question, a randomly generated set of numbers will be used so that students can see multiple iterations of the same question for extra practice. Please note that every set of numbers generated will still assess the same level of difficulty and objective.
- Students receive real-time feedback on every question answered.
- Practice mode has the same variety of question types included in all the PMT assignments including entering expressions and equations, graphing lines, fill-in-the-blank, and short response.
- Learning Aids are available including guided examples, step-by-step solutions, and video tutorials.
- Teachers also get access to preview the content students see.
- Student activity is meant to be for personal practice and is not included in teacher or student reports in *ThinkCentral* or *HMH Player*®.
- The PMT assignments are not adaptive.

Fluency Assignments

In addition to Personal Math Trainer® assignments, you have access to numerous digital exercises designed to support computational fluency tailored to each grade

Tracking and reporting on the fluency standards

In order to generate teacher and student reports from the assignments, log into the *ThinkCentral* dashboard and select the "PMT and Fluency Assignment" icon in the Resources section. Select "Fluency Assignments" and then select "Assign" next to the content you want to assign. You can also assign Fluency assignments via the *HMH Player*®.

There are a variety of fluency assignments available for each grade:

- **Fluency Checks:** Designed to assess grade-level fluency standards

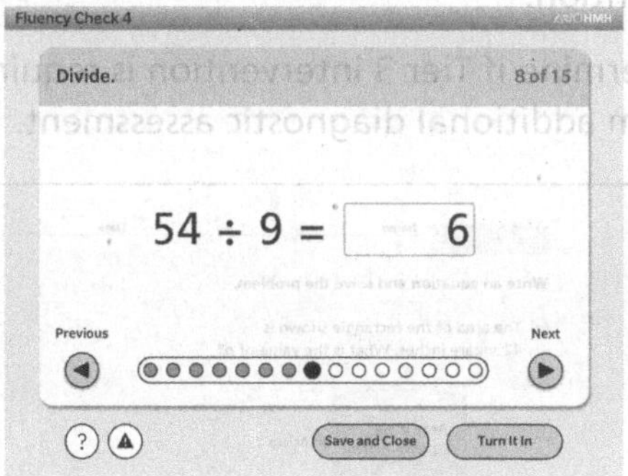

- **Grade-Level Fluency Builder:** Designed for practicing grade-level fluency standards
- **Future Fluency Starter:** Designed to prepare for next grade's fluency standards
- **Fluency Maintenance:** Designed for cumulative practice including fluency standards from previous grades.

Results of the assignments will be reflected on a variety of reports including standards-based reports available in *ThinkCentral* and *HMH Player*®.

Unlimited Fluency Practice via the Math Activity Center

Students also have unlimited, any-time access to Fluency assignments for practice via the online Math Activity Center. Each fluency question is algorithmically generated; each time a student accesses a question, a randomly generated set of numbers will be used so that students can see multiple iterations of the same question for extra practice. Please note that every set of numbers generated will still assess the same level of difficulty and objective. Student activity in the online Math Activity Center is meant for personal practice and is not included in teacher or student reports in *ThinkCentral* or *HMH Player*®. Teachers also get access to the online Math Activity Center to preview the content students see.

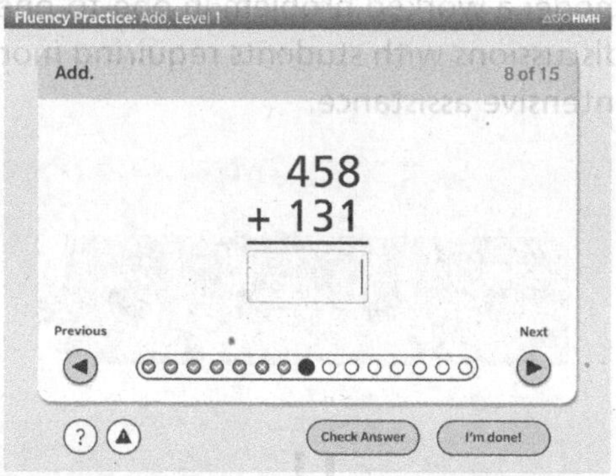

Using Response to Intervention

Real-Time Assessments and Intervention

Formative and diagnostic assessment and appropriate intervention activities have been incorporated into the online Math Activity Center.

RtI Support at Every Level

Comprehensive RtI support improves learning, and *Math Expressions* K–6 provides options for intervention at each tier. Systemic instruction provides explicit modeling of mathematical strategies and processes to strengthen essential skills.

- Tier 1 interactive activities provide scaffolded instruction in multiple modalities with explicit instruction of on-level content.
- Tier 2 interactive activities provide scaffolded instruction in multiple modalities for pre-requisite skills.
- Teacher-led Tier 3 downloadable PDF activities provide teacher talking points to model a worked problem in one-to-one discussions with students requiring more intensive assistance.

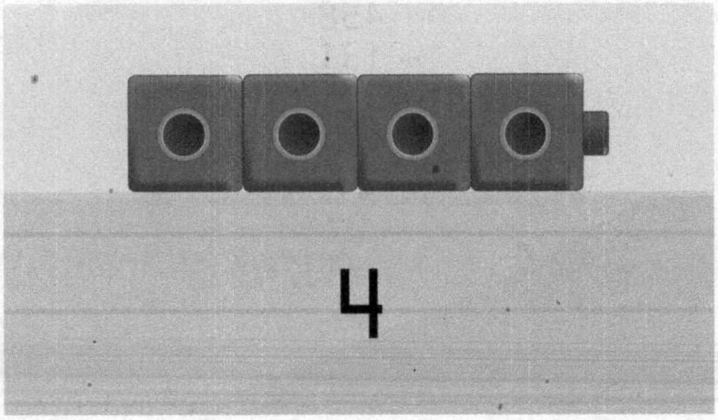

Formative

Use the Quick Quizzes to monitor and assess student learning and understanding, and adjust instruction accordingly.

Quick Quizzes are available in the Assessment Guide and in the PMT. Administer assessment at the completion of each Big Idea.

If a student answers one or two questions incorrectly, assign an interactive activity for Tier 1 intervention.

If a student answers more than two questions incorrectly, assign the interactive activity for Tier 2 intervention.

To determine if Tier 3 intervention is required, perform additional diagnostic assessment.

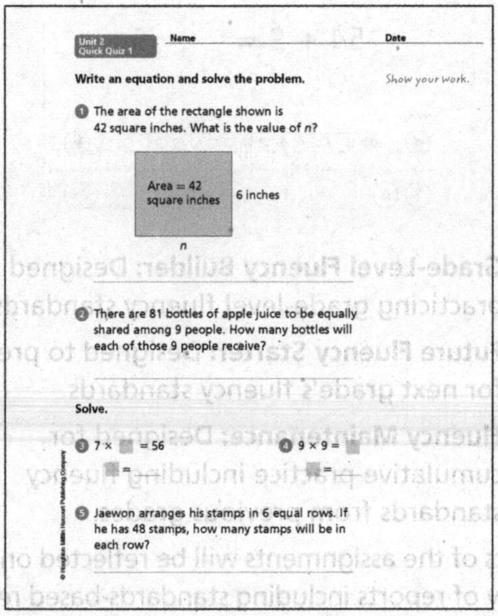
Unit 2 Quick Quiz 1 Name Date

Write an equation and solve the problem. *Show your work.*

1. The area of the rectangle shown is 42 square inches. What is the value of n?

2. There are 81 bottles of apple juice to be equally shared among 9 people. How many bottles will each of those 9 people receive?

Solve.

3. 7 × ■ = 56 ■ =
4. 9 × 9 = ■ ■ =

5. Jaewon arranges his stamps in 6 equal rows. If he has 48 stamps, how many stamps will be in each row?

Using Response to Intervention *(continued)*

Diagnostic

Tier 3 intervention is available online and incorporates a diagnostic approach through one-on-one teacher interviews and discussion. The Tier 3 assessment and activities provide:

- quick task-based evaluation of critical skills.
- response evaluation with intervention recommendations.
- worked problems with discussion tips.

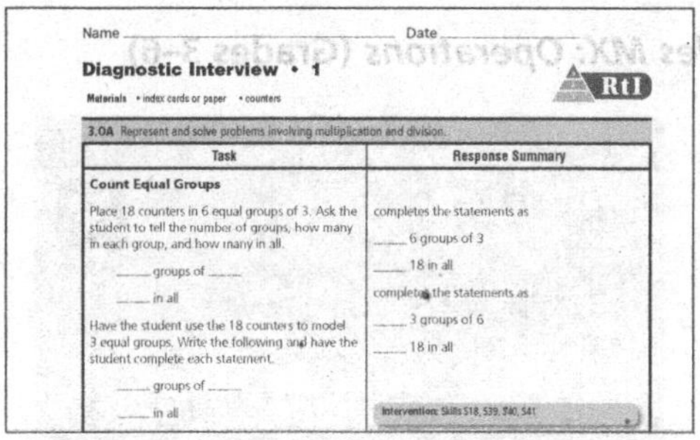

Name ______ Date ______

Diagnostic Interview • 1

Materials • index cards or paper • counters

RtI

3.OA Represent and solve problems involving multiplication and division.

Task	Response Summary
Count Equal Groups	
Place 18 counters in 6 equal groups of 3. Ask the student to tell the number of groups, how many in each group, and how many in all. ____ groups of ____ ____ in all	completes the statements as ____ 6 groups of 3 ____ 18 in all
Have the student use the 18 counters to model 3 equal groups. Write the following and have the student complete each statement. ____ groups of ____ ____ in all	completes the statements as ____ 3 groups of 6 ____ 18 in all Intervention: Skills S18, S39, S40, S41

Progression

Students requiring Tier 3 intervention may also benefit from the interactive experience of Tier 2 RtI.

Assign students completing a Tier 2 assignment the corresponding Tier 1 activity to follow through from pre-requisite skills to on-level content practice.

Online Interactive Games

The Online Math Activity Center features *Poggles MX*, a suite of educational video games. *Poggles MX* uses the engagement of video games to give students an immersive and fun way to practice math.

The Games

Our first edition of *Poggles MX* features four games. There is one game at the primary level (K–2) and three at the intermediate level (3–6).

Poggles MX: Addition and Subtraction **(Grades K–2)**

Students control a variety of yellow Poggles, and the only way to defeat the blue Poggles is by matching their value. Students practice one-to-one correspondence; 5-, 10-, and 100-groups; place value; skip counting; and multiples of 10. Many levels feature blue Poggles lined up as 2- and 3-digit numbers and require students to add or subtract them. In those levels, students can click the Poggles to see them regroup.

Poggles MX: Operations **(Grades 3–6)**

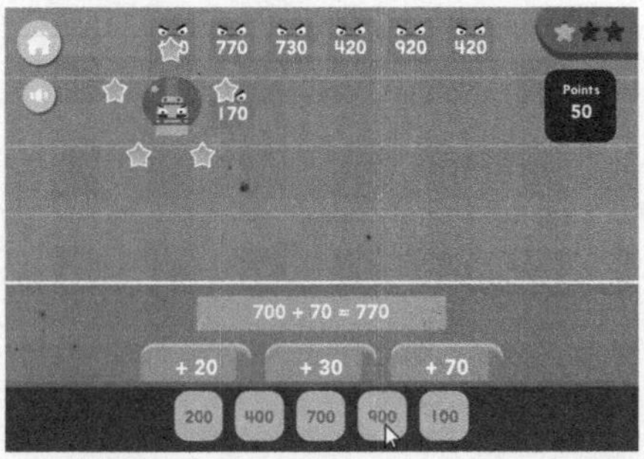

Blue Poggles are advancing towards the end zone, and the only way to stop them is by sending equivalent yellow Poggles their way. This time-based game helps students build fluency with basic operations, see patterns among operations, and develop facility with a wide range of *Math Expressions topics.*

Poggles MX: The Real Numbers **(Grades 3–6)**

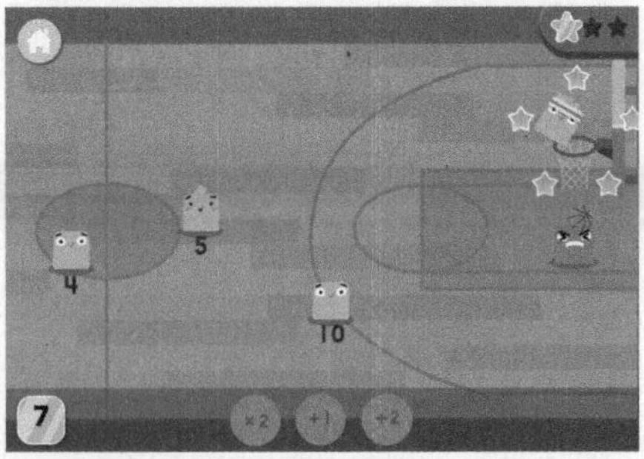

This basketball game gives students a starting value, a target value, and a set of basketballs to help them get there. This untimed puzzle game helps students develop facility with real number calculations and encourages them to think several steps ahead. Levels include decimals, fractions, mixed numbers, powers of 10, and more.

Poggles MX: Number Relationships **(Grades 3–6)**

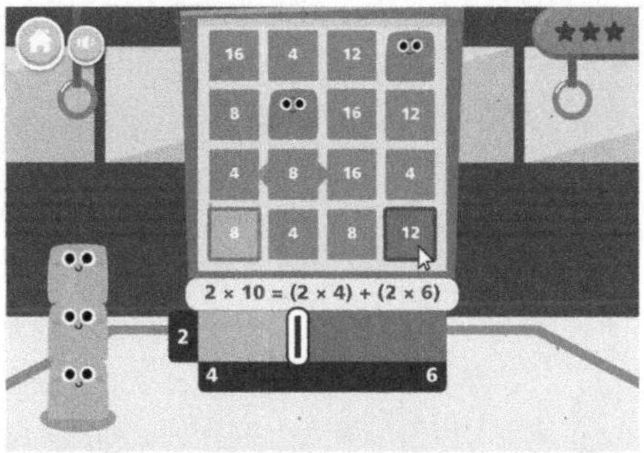

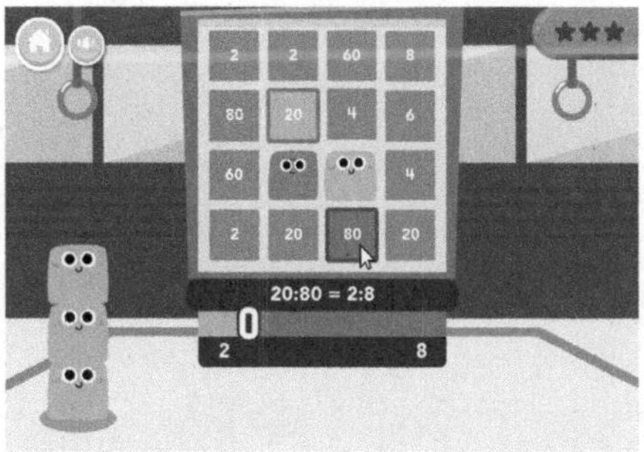

In this gymnastics game, students match pairs of numbers in a grid to pairs of numbers in a rectangle below the grid. In grades 3–5, the rectangle is typically divided into two parts, and the students look for pairs equal to the area of each part. In grade 6, the rectangle is typically divided into two numbers, and students look for pairs that correspond to an equivalent ratio.

Poggles MX in the Classroom

The thrill of completing a level, the low cost of making a wrong move, and the hundreds of innovative levels may help engage both struggling and advanced students alike. We suggest two ways of incorporating the games into your teaching.

- **Have students play using freeplay mode.**

All students can play the games for their grade freely, starting with Level 1. Students unlock levels sequentially, even if they unlock content not yet covered in class. They can also always return to previous levels. Intermediate-grade students may do this to achieve a new high score or earn more stars on a previously beat level. We recommend having students use freeplay mode when you are looking for productive ways for some students to pass time while you work with others.

- **Assign levels to students as classwork or homework.**

Teachers can assign individual levels as classwork or homework. A correlation chart is available online using the Games link in the Math Activity Center. We recommend using the games as a way to build numerical fluency skills. For example, if a student is struggling to multiply with 6, a game that practices the 6s count bys can be a fun way to practice. We also recommend using the games as an additional resource for practicing any lesson that has a matching game level. We suggest you play a game level yourself before assigning it. All levels are automatically unlocked for teachers.

Board Games

Board Games engage students to practice essential math skills in an enjoyable manner. In addition to the unit-based *Board Games*, students play the *Vocabulary Game* with vocabulary cards (found in each unit of the *Student Activity Book*). The *Vocabulary Game* gives students practice with each unit's vocabulary. Students will want to play these games several times, allowing them ample hands-on practice and mastery of essential skills.

Teacher Notes for *Board Games* include:

- objectives for each game
- materials for the game
- game instructions

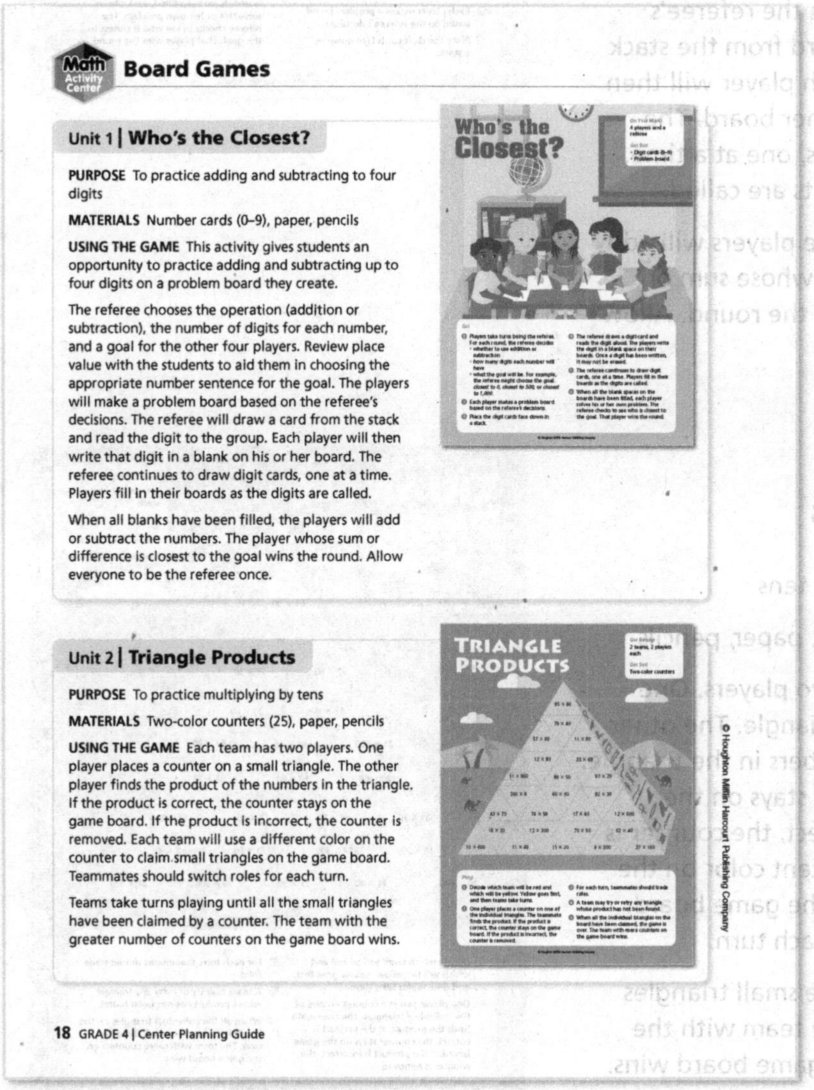

Board Games

Unit 1 | Who's the Closest?

PURPOSE To practice adding and subtracting to four digits

MATERIALS Number cards (0–9), paper, pencils

USING THE GAME This activity gives students an opportunity to practice adding and subtracting up to four digits on a problem board they create.

The referee chooses the operation (addition or subtraction), the number of digits for each number, and a goal for the other four players. Review place value with the students to aid them in choosing the appropriate number sentence for the goal. The players will make a problem board based on the referee's decisions. The referee will draw a card from the stack and read the digit to the group. Each player will then write that digit in a blank on his or her board. The referee continues to draw digit cards, one at a time. Players fill in their boards as the digits are called.

When all blanks have been filled, the players will add or subtract the numbers. The player whose sum or difference is closest to the goal wins the round. Allow everyone to be the referee once.

Unit 2 | Triangle Products

PURPOSE To practice multiplying by tens

MATERIALS Two-color counters (25), paper, pencils

USING THE GAME Each team has two players. One player places a counter on a small triangle. The other player finds the product of the numbers in the triangle. If the product is correct, the counter stays on the game board. If the product is incorrect, the counter is removed. Each team will use a different color on the counter to claim small triangles on the game board. Teammates should switch roles for each turn.

Teams take turns playing until all the small triangles have been claimed by a counter. The team with the greater number of counters on the game board wins.

18 GRADE 4 | Center Planning Guide

Board Games

Unit 1 | Who's the Closest?

PURPOSE To practice adding and subtracting to four digits

MATERIALS Number cards (0–9), paper, pencils

USING THE GAME This activity gives students an opportunity to practice adding and subtracting up to four digits on a problem board they create.

The referee chooses the operation (addition or subtraction), the number of digits for each number, and a goal for the other four players. Review place value with the students to aid them in choosing the appropriate number sentence for the goal. The players will make a problem board based on the referee's decisions. The referee will draw a card from the stack and read the digit to the group. Each player will then write that digit in a blank on his or her board. The referee continues to draw digit cards, one at a time. Players fill in their boards as the digits are called.

When all blanks have been filled, the players will add or subtract the numbers. The player whose sum or difference is closest to the goal wins the round. Allow everyone to be the referee once.

Unit 2 | Triangle Products

PURPOSE To practice multiplying by tens

MATERIALS Two-color counters (25), paper, pencils

USING THE GAME Each team has two players. One player places a counter on a small triangle. The other player finds the product of the numbers in the triangle. If the product is correct, the counter stays on the game board. If the product is incorrect, the counter is removed. Each team will use a different color on the counter to claim small triangles on the game board. Teammates should switch roles for each turn.

Teams take turns playing until all the small triangles have been claimed by a counter. The team with the greater number of counters on the game board wins.

Board Games *(continued)*

Unit 3 | Remainder or Not?

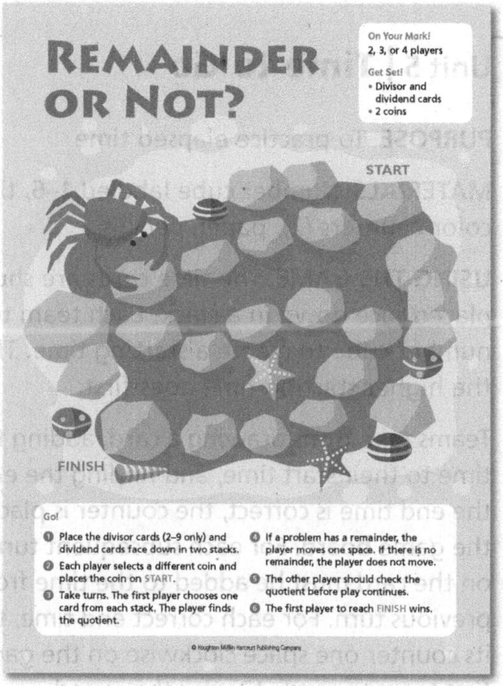

PURPOSE To practice mental math when dividing

MATERIALS Divisor and dividend cards, 2 coins

USING THE GAME The divisor and dividend cards should each be in a face down stack on the game board. Each player selects a different coin and places the coin on START. Players will alternate drawing one card from each stack and finding the quotient. If the quotient has a remainder, the player moves his or her coin one space. If there is not a remainder, then the player's coin does not move. The players should check each other's quotients.

The first player to reach FINISH wins.

Unit 4 | Factor Farm

PURPOSE To practice determining factors of whole numbers

MATERIALS Number cards (2–10), two-color counters (25)

USING THE GAME The number cards are shuffled and placed face down in a stack. Players take turns drawing a number card, determining if the number drawn is a factor of one of the numbers on the game board grid, and placing a counter on the number on the grid if it is a factor. Only one counter may be placed each turn.

The first player to place five counters on the grid in a horizontal, vertical, or diagonal line wins.

Have students create their own 5 × 5 grid and write a number from 2 to 40 in each space. Numbers should only be used once. Then have students play again.

Unit 5 | Time to Go

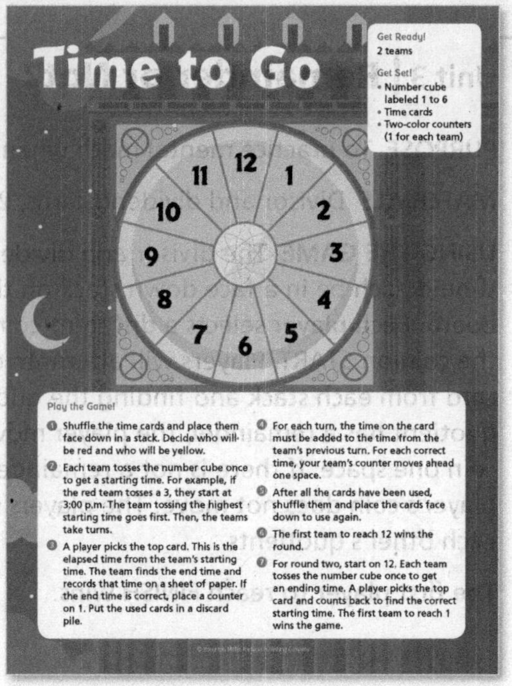

PURPOSE To practice elapsed time

MATERIALS Number cube labeled 1–6, time cards, two-color counters (2), paper, pencils

USING THE GAME The time cards are shuffled and placed face down in a stack. Each team tosses the number cube to create a starting time. The team with the higher starting time goes first.

Teams take turns drawing a card, adding the elapsed time to their start time, and finding the end times. If the end time is correct, the counter is placed on 1 on the game board. For each subsequent turn, the time on the card must be added to the time from the team's previous turn. For each correct end time, the team moves its counter one space clockwise on the game board. The first team to reach 12 wins the round.

For the next round, play the game again with teams' counters beginning on 12 and start times of 12:00 p.m. In this round, when cards are drawn the teams will count back to find the start time. For each correct start time, the team moves its counter one space counter-clockwise on the game board. The first team to reach 1 wins the round.

Unit 6 | Fraction Action

PURPOSE To practice identifying fractions

MATERIALS Fraction domino cards

USING THE GAME The fraction domino cards are shuffled and placed face down in a stack. Each player draws five cards to hold in his or her hands. Then each player places the top card of the stack face up in the playing area. Players will alternate turns by matching a picture with its fraction or a fraction with its picture. If you need to, take cards from the stack until you get a fraction card that can be played.

Play continues until the stack of cards is gone. The player who has the least number of cards in his or her hands when the stack is gone wins the game.

Unit 7 | Order, Please!

PURPOSE To practice ordering decimals

MATERIALS Set of index cards with these decimals: 0.1–0.9, 0.15–0.25, 0.01–0.09

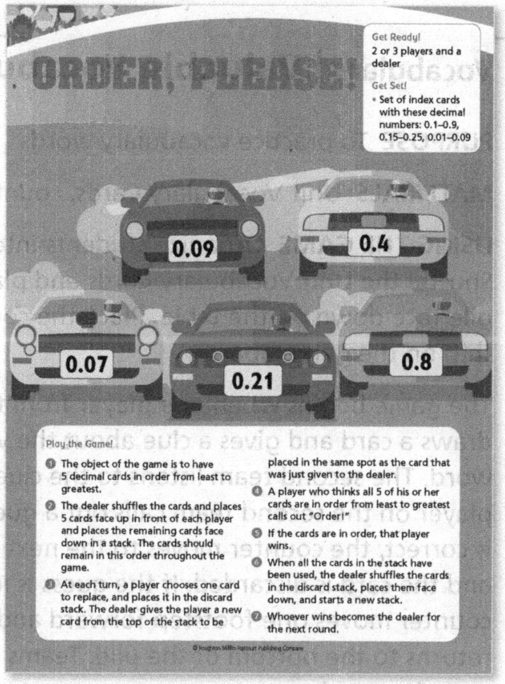

USING THE GAME Cut 15 index cards in half and write the decimals on each halved card. The dealer shuffles the cards and then places five cards face up in front of each player, including him- or herself. The cards stay in the order they are dealt for the whole game. The dealer then places the remaining cards face down in a stack between the players. In turn, each player picks a card from his or her five cards and discards it. The dealer then gives that player a card from the stack, which is placed in the position of the discarded card. Players continue to switch cards until one player has five cards in order from least to greatest, at which point he or she calls out "Order!"

The first player to have five cards in order from least to greatest wins the game. For variation, the players could order the cards from greatest to least.

Unit 8 | What's Your Angle?

PURPOSE To practice drawing and measuring angles

MATERIALS For partners: protractor pattern or protractor, paper, pencil

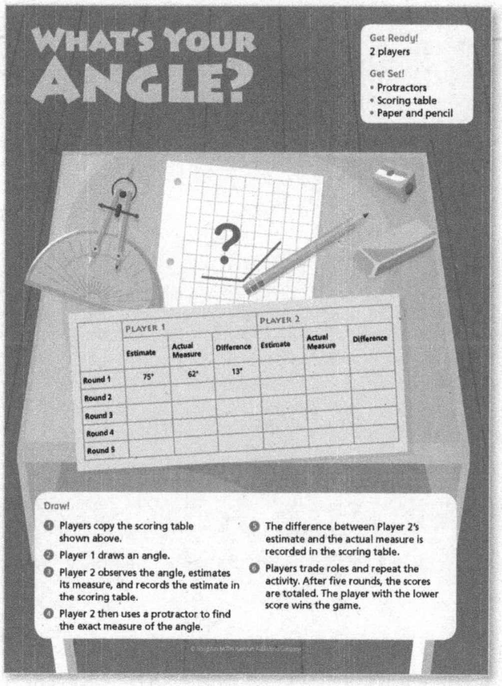

USING THE GAME Prior to beginning the activity, partners create a scoring table for recording their measurements.

The game begins as one player draws an angle. The second player observes the drawing and predicts the angle's measure. Using a protractor, the second player measures and records the actual measure. Then the second player records the difference between the actual measure and the predicted measure.

Players trade roles and repeat the process to finish the round. After five rounds, the scores are totaled by adding the differences from each round. The player with the lower score wins the game.

Vocabulary | Valuable Vocabulary

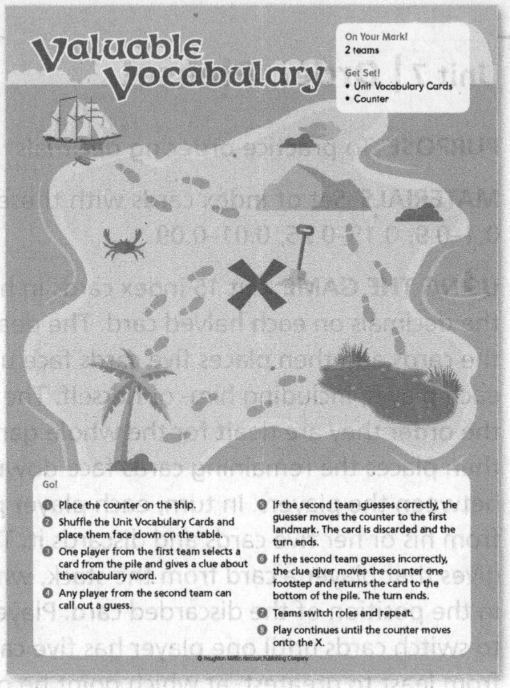

PURPOSE To practice vocabulary words

MATERIALS Unit Vocabulary Cards, counter

USING THE GAME Organize students into two teams. Shuffle the Unit Vocabulary Cards and place them in a pile face down on the table. Place the counter on the ship on the game board.

The game begins when one player from the first team draws a card and gives a clue about the vocabulary word. The second team listens to the clue, and one player on the second team calls out a guess. If the guess is correct, the counter moves to the next landmark and the card is discarded. If the guess is incorrect, the counter moves one footstep forward and the card returns to the bottom of the pile. Teams switch roles and play continues.

The game continues until the counter is moved onto the X on the game board.

Number Cards

0	1	2	3
4	5	6	7
8	9	10	11
12	13	14	15
16	17	18	19

Remainder or Not?

Dividend Cards

110	180	200	230	250
310	340	370	400	420
460	490	550	580	610
650	700	760	830	980

Divisor Cards

2	3	4	5	6
7	8	9	10	20
30	40	50	60	70
80	90	20	40	50

Number Cube Pattern

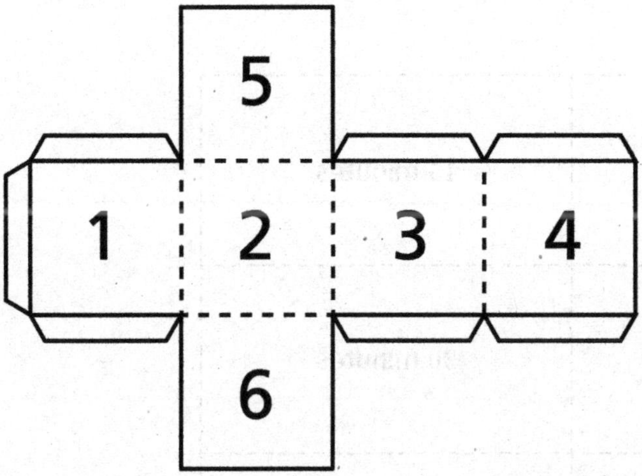

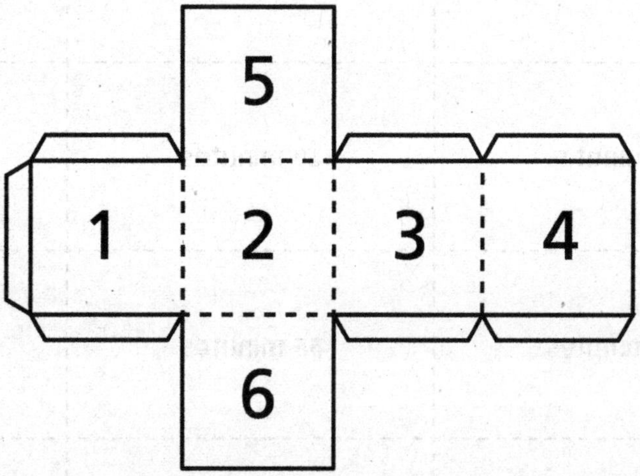

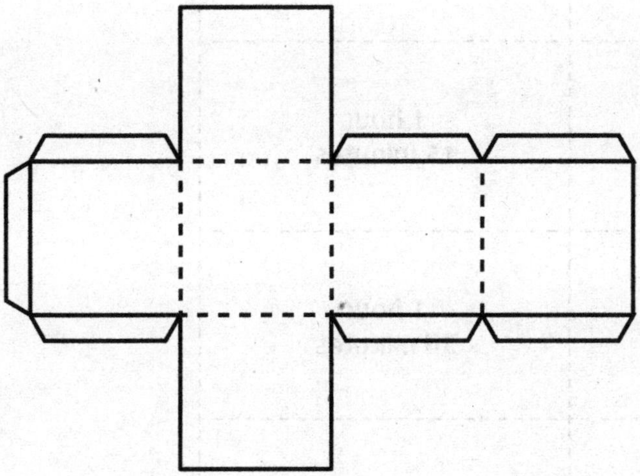

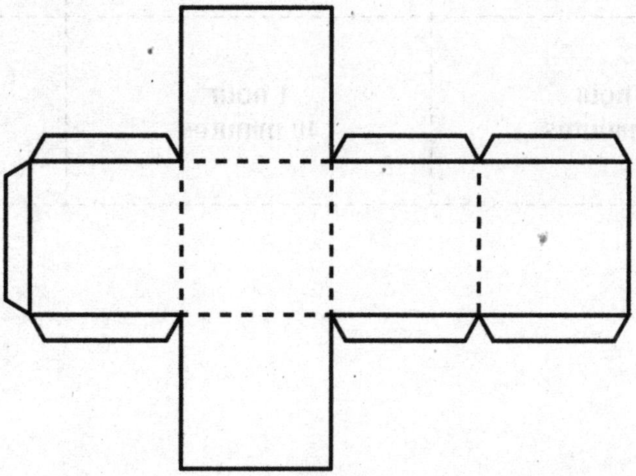

Time to Go

Time Cards

5 minutes	10 minutes	15 minutes
20 minutes	25 minutes	30 minutes
35 minutes	40 minutes	45 minutes
50 minutes	55 minutes	1 hour
1 hour 5 minutes	1 hour 10 minutes	1 hour 15 minutes
1 hour 20 minutes	1 hour 25 minutes	1 hour 30 minutes
1 hour 35 minutes	1 hour 40 minutes	1 hour 45 minutes

Fraction Action

Domino Cards

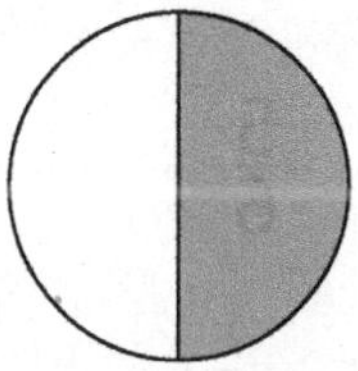	$\frac{1}{5}$	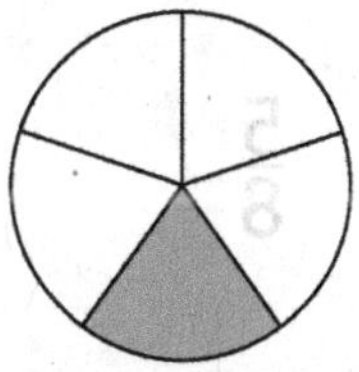	$\frac{3}{6}$
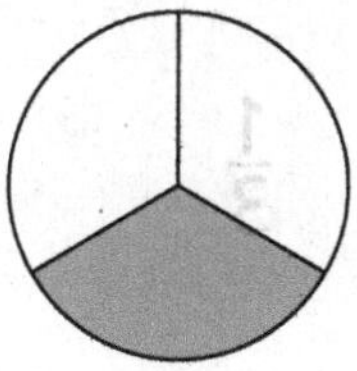	$\frac{6}{8}$	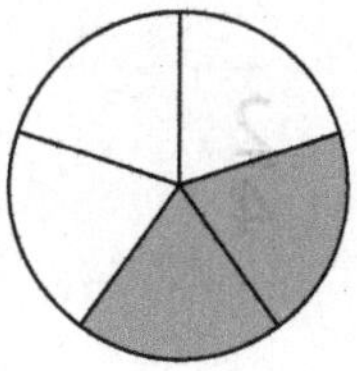	$\frac{3}{8}$
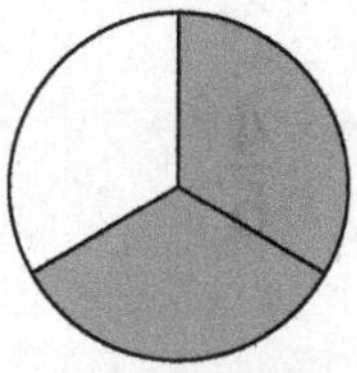	$\frac{1}{8}$	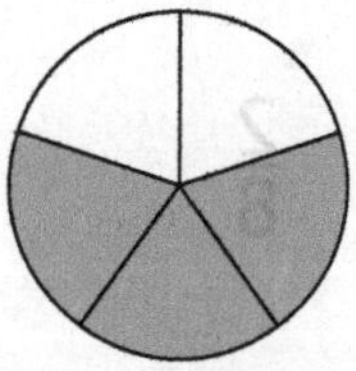	$\frac{4}{8}$
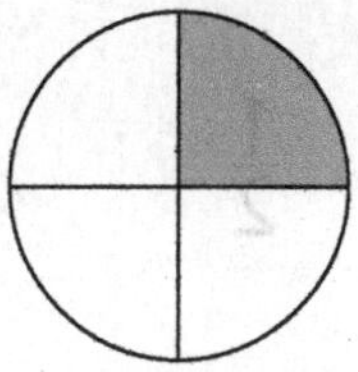	$\frac{7}{8}$	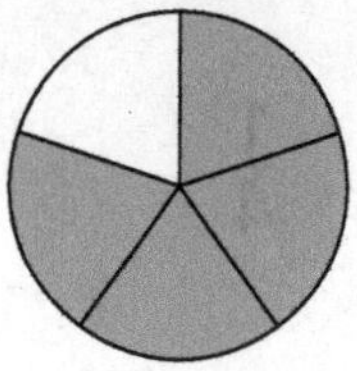	$\frac{2}{6}$
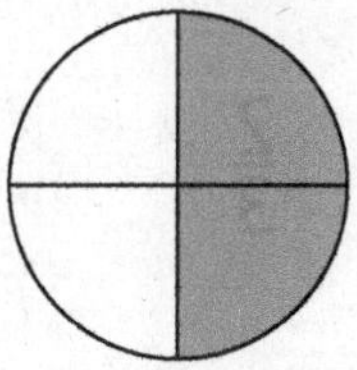	$\frac{1}{6}$	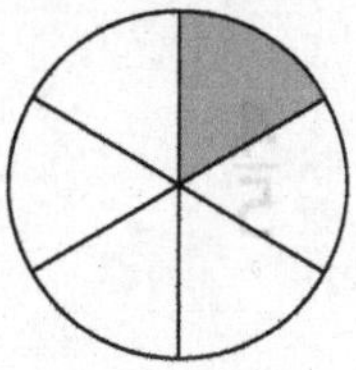	$\frac{3}{5}$
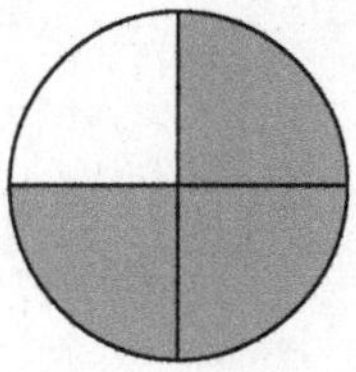	$\frac{2}{5}$	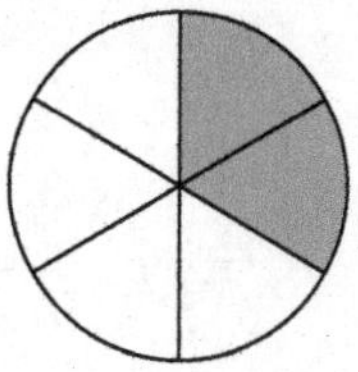	$\frac{3}{4}$

Fraction Action

Domino Cards

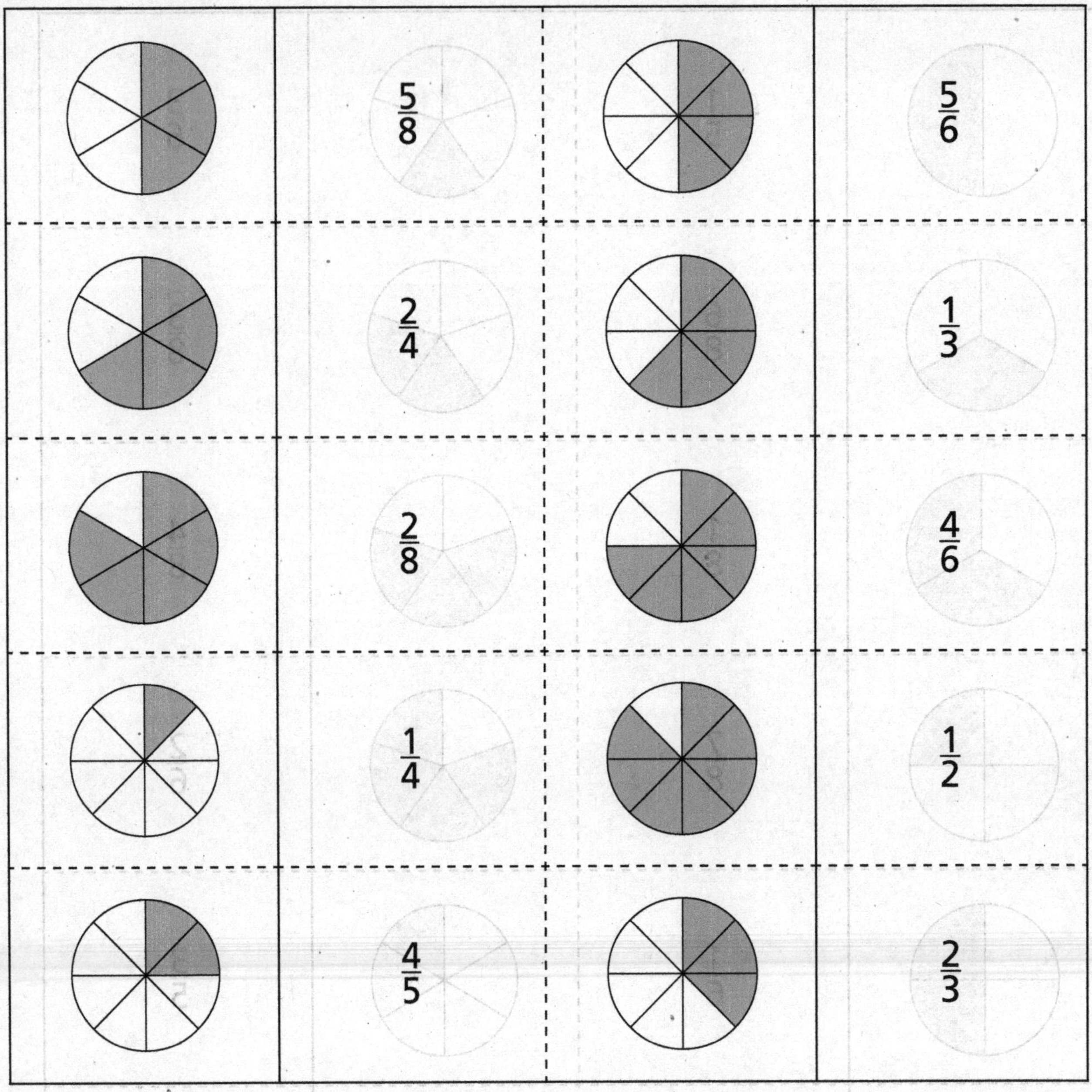

Protractor Pattern

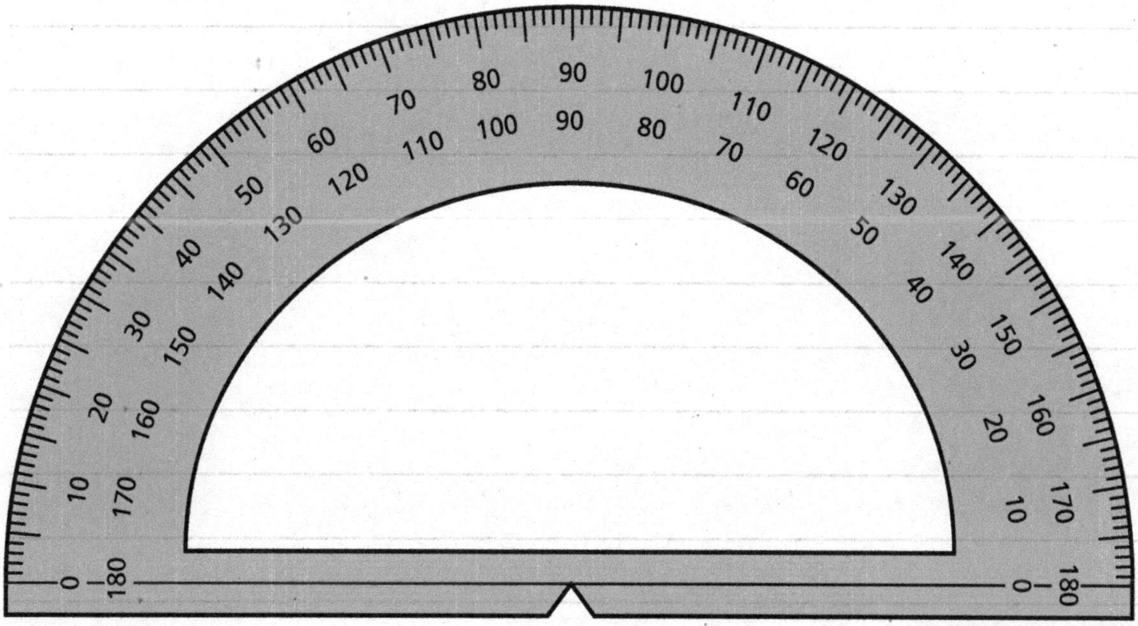

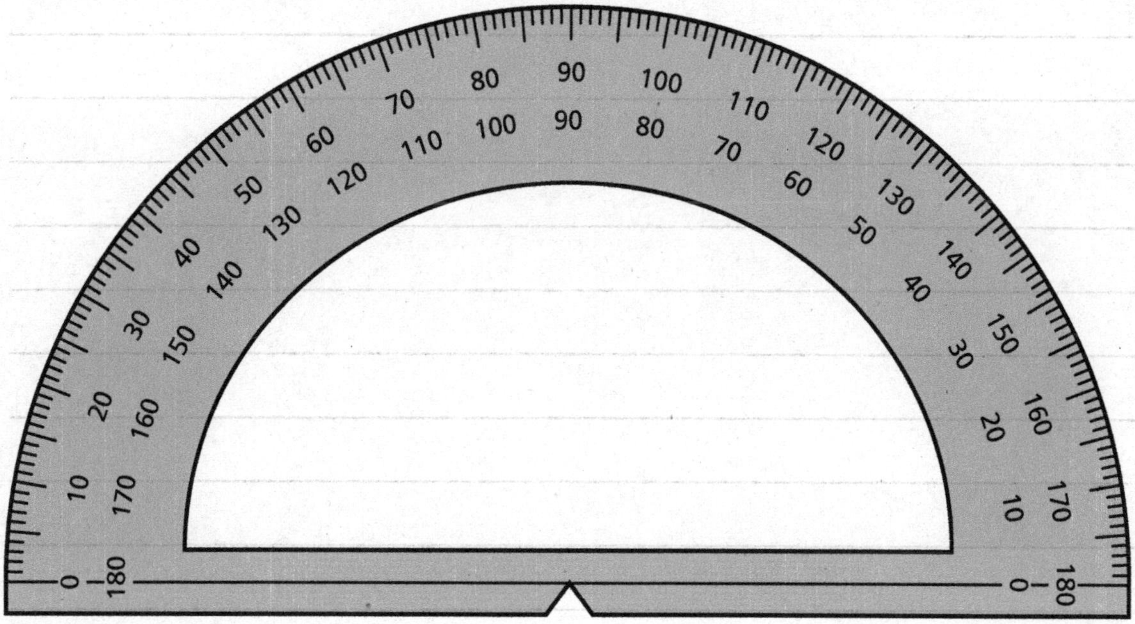

Notes:

In-Depth Inquiry Based Tasks

In-Depth Inquiry Based Tasks are open-ended unit activities that help you reinforce or extend mathematical skills while also increasing student engagement in learning. Through drawings, conceptual language, inquiry, and peer collaboration, the tasks help students apply mathematics to real-world scenarios. There is a different *In-Depth Inquiry Based Task* for every unit. At the end of each unit, students will have mastery of skills necessary to complete the unit activity. To challenge students performing above level, these activities can be utilized mid-unit. Students will engage, explore, explain, and extend mathematical concepts with each activity. The open-ended format puts students at the helm of their learning, while still providing a "toolbox" of supports that offers mathematical guidance when needed.

Teacher Notes for *In-Depth Inquiry Based Tasks* include:

- objectives for each activity
- materials needed for each activity
- time needed for each activity
- questions to consider before, during, and/or after the activity

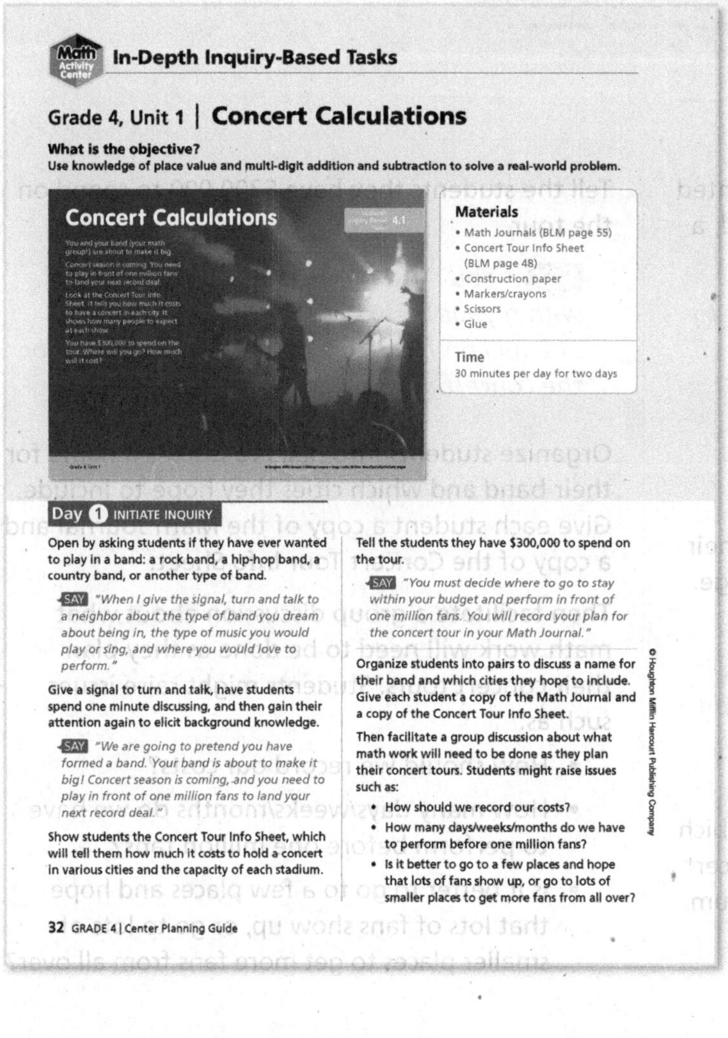
In-Depth Inquiry-Based Tasks

Grade 4, Unit 1 | **Concert Calculations**

What is the objective?
Use knowledge of place value and multi-digit addition and subtraction to solve a real-world problem.

Materials
- Math Journals (BLM page 55)
- Concert Tour Info Sheet (BLM page 48)
- Construction paper
- Markers/crayons
- Scissors
- Glue

Time
30 minutes per day for two days

Day 1 INITIATE INQUIRY

Open by asking students if they have ever wanted to play in a band: a rock band, a hip-hop band, a country band, or another type of band.

SAY *"When I give the signal, turn and talk to a neighbor about the type of band you dream about being in, the type of music you would play or sing, and where you would love to perform."*

Give a signal to turn and talk, have students spend one minute discussing, and then gain their attention again to elicit background knowledge.

SAY *"We are going to pretend you have formed a band. Your band is about to make it big! Concert season is coming, and you need to play in front of one million fans to land your next record deal."*

Show students the Concert Tour Info Sheet, which will tell them how much it costs to hold a concert in various cities and the capacity of each stadium.

Tell the students they have $300,000 to spend on the tour.

SAY *"You must decide where to go to stay within your budget and perform in front of one million fans. You will record your plan for the concert tour in your Math Journal."*

Organize students into pairs to discuss a name for their band and which cities they hope to include. Give each student a copy of the Math Journal and a copy of the Concert Tour Info Sheet.

Then facilitate a group discussion about what math work will need to be done as they plan their concert tours. Students might raise issues such as:

- How should we record our costs?
- How many days/weeks/months do we have to perform before one million fans?
- Is it better to go to a few places and hope that lots of fans show up, or go to lots of smaller places to get more fans from all over?

32 GRADE 4 | Center Planning Guide

Grade 4, Unit 1 | Concert Calculations

What is the objective?
Use knowledge of place value and multi-digit addition and subtraction to solve a real-world problem.

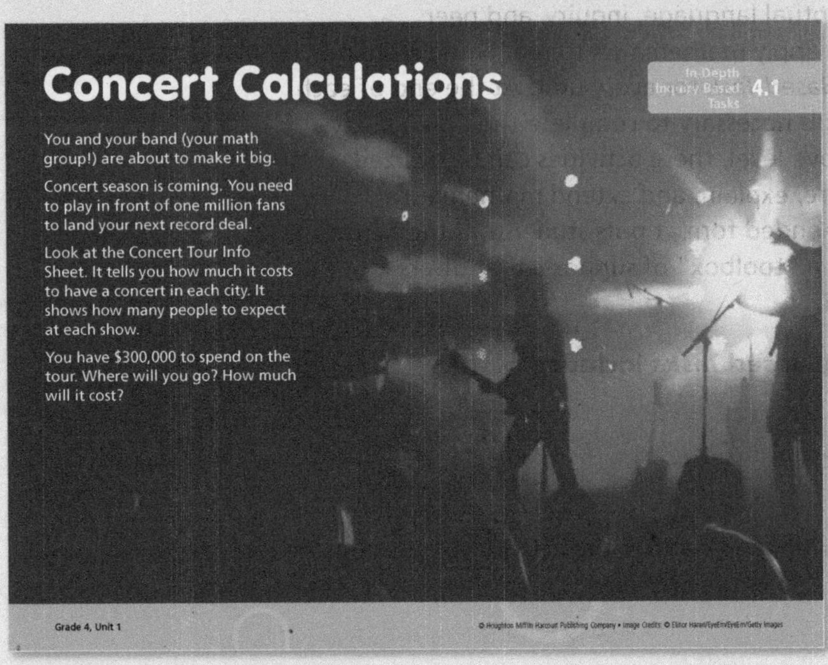

Materials

- Math Journals (BLM page 55)
- Concert Tour Info Sheet (BLM page 48)
- Construction paper
- Markers/crayons
- Scissors
- Glue

Time

30 minutes per day for two days

Day 1 INITIATE INQUIRY

Open by asking students if they have ever wanted to play in a band: a rock band, a hip-hop band, a country band, or another type of band.

SAY *"When I give the signal, turn and talk to a neighbor about the type of band you dream about being in, the type of music you would play or sing, and where you would love to perform."*

Give a signal to turn and talk, have students spend one minute discussing, and then gain their attention again to elicit background knowledge.

SAY *"We are going to pretend you have formed a band. Your band is about to make it big! Concert season is coming, and you need to play in front of one million fans to land your next record deal."*

Show students the Concert Tour Info Sheet, which will tell them how much it costs to hold a concert in various cities and the capacity of each stadium.

Tell the students they have $300,000 to spend on the tour.

SAY *"You must decide where to go to stay within your budget and perform in front of one million fans. You will record your plan for the concert tour in your Math Journal."*

Organize students into pairs to discuss a name for their band and which cities they hope to include. Give each student a copy of the Math Journal and a copy of the Concert Tour Info Sheet.

Then facilitate a group discussion about what math work will need to be done as they plan their concert tours. Students might raise issues such as:

- How should we record our costs?
- How many days/weeks/months do we have to perform before one million fans?
- Is it better to go to a few places and hope that lots of fans show up, or go to lots of smaller places to get more fans from all over?

Day 1 (continued)

Encourage students to determine the answers to their questions together.

SAY *"You must keep a record of your costs and number of fans in your Math Journal. You can use any of the addition and subtraction models learned in Unit 1 to show how much of your budget you have used, how much is left, how many fans you have planned to play for, and how many more fans you need to reach one million."*

Have students model two or three stops on a model concert tour, ways to record their spending, and number of fans attending in their journals.

Day 2 COMPLETE THE TASK

Gather the group and remind the students of Day 1's discussion. Distribute Math Journals again.

Have pairs begin planning their concert tours. Remind students to record their planning and math thinking in their journals clearly. Act as a tour manager, answering questions and suggesting math models to use when students feel stuck. For example, as a way to record their calculations, students might want to use the New Groups Above Method of notation in their journals:

Add the ones:
7 ones + 7 ones = 14 ones
14 ones = 10 ones + 4 ones
Group 10 ones as 1 ten.

$$\begin{array}{r} {\scriptstyle 1} \\ 66{,}387 \\ +78{,}297 \\ \hline 4 \end{array}$$

or

- Ungroup 1 ten:
  5 tens + 4 ones =
  4 tens + 14 ones
- Ungroup 1 thousand:
  9 thousands + 0 hundreds =
  8 thousands + 10 hundreds

$$\begin{array}{r} {\scriptstyle 8\ \ 10\ 4\ 14} \\ 6\not{9}{,}\not{0}\not{5}\not{4} \\ -43{,}128 \\ \hline 25{,}926 \end{array}$$

Or students might prefer to model subtraction using the "adding up" method:

$$\begin{array}{r} 787{,}136 \\ -138{,}258 \\ \hline 648{,}878 \\ {\scriptstyle 1\,1\ \ 1\,1} \end{array}$$

If students finish their concert tour plan early, they could design a poster for their band that lists all the tour stops and stadium capacities.

Once all the pairs are done, discuss the task and ask students to share what they recorded in their journals. Possible questions might include:

- I see that many of you still have money left in your budget. Why? How much? Is it enough to make one more stop?
- Were any of you unable to land your next record deal? Explain.
- Was there any advantage to visiting larger cities over smaller cities or vice versa?
- Explain the math you used to meet the goal of one million while staying within your budget.

SAY *"When authors go on book tours, they must make similar decisions. Is it better for them to visit big-city bookstores, where authors draw large crowds and can sell more books, or smaller cities, which have fewer fans but are less expensive to visit? Plan an author's book tour in your Math Journal."*

Grade 4, Unit 2 | The Zoo and You

What is the objective?

Use knowledge of multi-digit addition, subtraction, and multiplication to solve a real-world problem.

Materials

- Math Journals (BLM page 55)
- Zoo Trip Planner (BLM page 49)

Time

30 minutes per day for two days

Day 1 INITIATE INQUIRY

Open by asking students if they have ever gone to the zoo or on a field trip.

SAY *"When I give the signal, turn and talk to a neighbor about a time you went to a zoo or on a field trip."*

Give a signal to turn and talk, have students spend one minute discussing, and then gain their attention again to elicit background knowledge. Announce that the group will be planning a pretend field trip to the zoo.

SAY *"Think about what needs to be paid for when a class goes on a field trip."*

Show students the Zoo Trip Planner that they will use to complete the activity.

SAY *"This tells you the cost of the bus, gasoline, lunches, and zoo admission."*

Give the students a budget for the trip. The teacher will want to calculate this ahead of time based on the number of students in the class. If the teacher opts to include the challenge activity, he or she should set a budget that is lower than the cost the field trip will be, forcing the students to plan the fundraiser. If the teacher opts to omit the challenge activity, he or she should set a budget that is slightly higher than the cost the field trip will be, so there is no need for a fundraiser.

Organize students into pairs to discuss the field trip and plan ways to track their costs and stay within budget.

Then facilitate a whole-class discussion about what math work will need to be done for the task.

SAY *"What do we need to think about ahead of time for our field trip to the zoo?"*

Students might raise issues such as:

- How will we record our costs and what we have left in our budget?

Day 1 (continued)

- How many students will be going on the trip? Chaperones? Teachers?
- How many buses will we need? How many people can ride on one bus?
- How many gallons of gas will the trip require?
- Can we pack our own lunches to save money?
- What happens if another class wants to go too? Will we have a greater budget?

Encourage students to determine the answers to their questions.

SAY *"You must keep a record of your costs in your Math Journal. You can use any of the multiplication models learned in Unit 2 to show how much money the buses, gasoline, lunches, and tickets will cost. Plus you can use any of the addition, subtraction, and multiplication models you learned about in Unit 2 when calculating your total costs and amount left in your budget."*

Give each student a copy of the Math Journal and each group a copy of the Zoo Trip Planner. Provide time for students to begin working on the scenario in their Math Journals.

Day 2 COMPLETE THE TASK

Gather the class and remind the students of the Zoo Field Trip scenario. Ask a few students to share their work so far.

Remind students to record their planning and math thinking in their Math Journals. For example, as a way to record the cost of the tickets to enter the zoo in their journals, students might want to use this method:

$$\begin{array}{r} 28 = 20 + 8 \\ \times\ 12 = 10 + 2 \\ \hline 10 \times 20 = \quad 200 \\ 10 \times 8 = \quad 80 \\ 2 \times 20 = \quad 40 \\ 2 \times 8 = \quad 16 \\ \hline 336 \end{array}$$

Or review adding and subtracting of multi-digit numbers to help students find total costs or what is left of their budgets:

$$\begin{array}{r} 216 \\ -\ 67 \\ \hline 149 \end{array}$$

Add the ones:
6 ones + 5 ones = 11 ones
11 ones = 10 ones + 1 ones
Group 10 ones as 1 ten.

$$\begin{array}{r} {}^{1}\\ 376 \\ +\ 15 \\ \hline 1 \end{array}$$

Once the field trips have been planned and budgets met, ask students to share what they recorded in their journals. Possible discussion questions might include:

- I see that many of you still have money left from your budget. Why?
- What might the class do if the trip costs more than our budget?
- After calculating the cost of the tickets, how did you know how much you had left to spend?
- What is the greatest expense? Could there be a way of making the field trip cost less?
- How did addition, subtraction, and multiplication help you solve this real-world problem?

SAY *"What if the school doesn't have money to pay for the field trip? Plan a popcorn sale that covers the costs. How much does it cost you to make each bag of popcorn? How much would you need to charge per bag? How many popcorn bags would you need to sell to pay for all students to go on the field trip?"*

Grade 4, Unit 3 | The Great Bead Divide

What is the objective?
Use knowledge of division with whole numbers to solve a real-world problem.

Materials
- Math Journals (BLM page 55)
- The Great Bead Divide (BLM page 50)
- Colored pencils

Time
25 minutes per day for two days

Day 1 INITIATE INQUIRY

Open by asking students if they have ever had a pen pal or know what a pen pal is.

SAY *"When I give the signal, turn and talk to a neighbor about pen pals."*

Give a signal to turn and talk, have students spend one minute discussing, and then gain their attention again to elicit background knowledge.

SAY *"Today we will pretend we have pen pals from De Colores Elementary School. "Colores" means "colors" in Spanish, so we want to send our pen pals a colorful gift. We will pretend to make beaded animal key chains for our friends to wear on their backpacks."*

Show the students the photos of the cow, lizard, and dog beaded key chains on the front and back of the In-Depth Inquiry Based Task card.

SAY *"You will use the information on this Great Bead Divide info sheet to find out how many of each type of animal you can make with the beads you have. Look at the beaded-animal patterns to find out how many beads you need of each color to make each animal. The info sheet also says how many beads you have of each color."*

Tell the students they will use Digit by Digit, Place Value Sections, or Expanded Notation to show their math thinking.

SAY *"The first step is to color the beaded animal patterns according to the color key on the info sheet."*

Provide time for students to color the bead patterns. Make sure their coloring matches the code, otherwise the numbers won't divide evenly.

SAY *"Now look at the number of beads you have in all. Turn and talk with a partner about how knowing the number of beads will help you find out how many beaded animals you can make."*

Give the students three minutes to discuss and begin trying to solve the problem.

Day 1 (continued)

Gain the group's attention again and discuss what sort of math work will need to be done before, during, and after the task. Students might ask:

- How will knowing the total number of beads help us find out how many beaded animals we can make?
- What if the same color bead, such as black, is used on two different animals?
- Can there be remainders, or not enough to go around equally? (The answer to this needs to be no.)
- Can there be extra beads left over? (The answer to this needs to be no.)
- Do the animals we make have to have the exact same number and color of beads in the pattern? (The answer to this needs to be yes.)

Encourage students to determine the answers to their questions, except for the questions that are followed by specific answers.

SAY *"You will need to show your division methods in your Math Journal. You can use any of the models learned in Unit 3."*

For example, as a way to record how they divide the beads, students might want to make arrays in their journals:

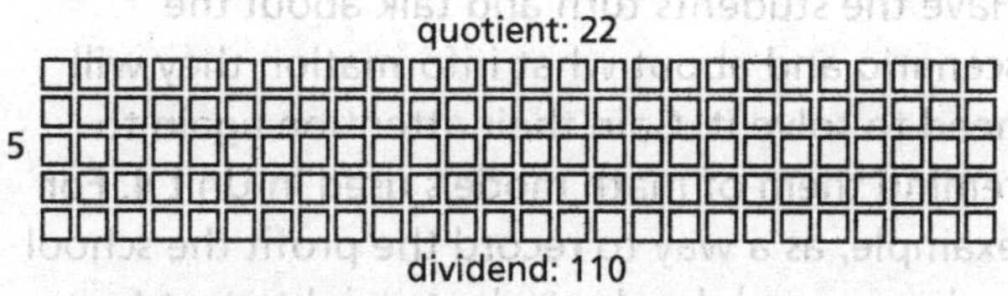

Or students might prefer to model the division using the digit-by-digit method:

$$\begin{array}{r} 26 \\ 5\overline{)130} \\ -10 \\ \hline 30 \\ -\ 30 \\ \hline 0 \end{array}$$

Distribute Math Journals and give the students time to work with a partner to begin solving the problem. Bring the group back together at the end so students can share examples of their work.

Day 2 COMPLETE THE TASK

Gather the group and remind the students of Day 1's discussion and main takeaways.

Instruct students to finish calculating how many beaded animals they can make based on the total number of beads provided. Remind students to show their math thinking and record their equations in their Math Journals.

Once the beaded-animal scenario has been solved, discuss the task and ask students to share what they recorded in their journals. Possible discussion questions might include:

- Did you have any difficulties dividing the beads equally? Why?
- Which animal was the most difficult to figure out? Why?
- What division methods did you use?
- What did you do when two or more beaded animals shared the same color beads?

SAY *"Imagine that your group had extra beads and had so much fun making beaded animals, you made 96 more animals than you needed for your pen pals. You give the extra animals to the 27 students in another fourth grade class. How many beaded animals will each student get? How many will be left over? Show how you know."*

Grade 4, Unit 4 | Buy the Book!

What is the objective?
Use knowledge of equations and word problems to solve a real-world problem.

Materials

- Math Journals (BLM page 55)
- Buy the Book Info Sheet (BLM page 51)
- Crayons or colored pencils (optional)

Time

40 minutes per day for two days

Day 1 INITIATE INQUIRY

Open by asking students if they have ever been to a book fair or discuss a book fair your school has hosted.

SAY *"When I give the signal, turn and talk to a neighbor about a time you went to a book fair and the type of books that were for sale."*

Give a signal to turn and talk, have students spend one minute discussing, and then gain their attention again to elicit background knowledge.

Announce that the groups will be planning a pretend book fair for Longfellow School.

SAY *"You must order enough books so that each child has a choice of at least five. Order books that children from all grade levels will like. Use the Buy the Book Info Sheet to learn more about the students at Longfellow."*

Tell the children it costs the school $2 to buy each book, and then the school sells each book for $6.

SAY *"You need to find out how much money the school will earn in all if every student buys three books."*

Have the students turn and talk about the scenario and about what information they will need to solve it. Gain their attention again to remind them of math models used in Unit 4. For example, as a way to record the profit the school makes on each book, students might want to make a break-apart drawing in their journals:

6
/\
2 p

$6 = 2 + p$ 2 and 4 make 6, so $p = 4$

Day 1 (continued)

Or students might prefer to use one equation with more than one step:

Write an equation for the whole problem.

Let n = the total profit

Price of book at fair.	Cost for school to buy the book.	# of books each students buys	# of Students buying books

$$(6 - 2) \times 3 \times \underline{153} = n$$

$$\underline{1{,}836} = n$$

Provide students with Math Journals and Buy the Book Info Sheets. Give them time to work with a partner to begin solving the problem and recording their thinking in the journals. Gain the group's attention again.

SAY *"Tomorrow you will learn more about the book fair at Longfellow School and solve another problem related to it."*

Day 2 COMPLETE THE TASK

Gather the group and remind the students of Day 1's work. Distribute the Math Journals and Buy the Book Info Sheet.

SAY *"Yesterday you began to find out how much money Longfellow School will earn from the book fair if each student buys three books. Today you will learn why the school is raising money."*

Tell the students that Longfellow needs $6,000 to buy new musical instruments.

SAY *"Does selling three books per student earn enough? How many more books would each student need to buy to reach that goal? You will keep a record of your equations and solutions in your Math Journal. You can use any of the reasoning models learned in Unit 4 to show how many books are needed, the cost, and the profit."*

Provide time for students to complete yesterday's problem and then determine how the school can reach its goal. Encourage students to use a bar graph to display the number of books the school needs to sell at each grade level to meet the goal.

Once the book fair planning task has run its course, ask students to share what they recorded in their journals. Possible discussion questions might include:

- How did you use your bar graph while planning the book fair?
- How did writing equations help you plan the book fair? Explain.
- How much profit would Longfellow School make at the book fair?
- Can you find the profit the book company makes? Explain.
- What equation could you write and solve to determine the profit if each student buys 6 books instead of just 3? Is there more than one equation that could represent this?
- How did rectangle models and break-apart drawings help you plan the book fair?
- Optional: Draw a map of the book-sale space. If you have 9 tables, how many books need to go on each table?

SAY *"Sometimes schools earn money through bake sales instead of book sales. What information would you need to solve a similar problem if you were selling muffins and granola bars instead of books?"*

Grade 4, Unit 5 | The Perfect Park

What is the objective?
Use knowledge of measurement to solve a real-world problem.

Materials
- Math Journals (BLM page 55)
- Centimeter Grid (BLM page 52)
- Ruler
- Tape (optional)

Time
40 minutes per day for two days

Day 1 INITIATE INQUIRY

Open by asking students if they ever go to neighborhood parks or have visited any special parks in other cities.

SAY *"When I give the signal, turn and talk to a neighbor about parks. Tell about the features you enjoy the most."*

Give a signal to turn and talk, have students spend one minute discussing, and then gain their attention again to elicit background knowledge. Announce that today every student gets to be a landscape architect, designing and drawing a plan for the park of their dreams.

SAY *"Your dream-park design must be 2,000 square meters and can include any park feature you want. Pretend that cost is no object, and let your imaginations run wild. Today, in your Math Journals, you will plan what to put in your park and how to record the lengths, perimeters, and areas of each feature. You also need to think about how to convert from meters to centimeters to draw your park to scale. Tomorrow you will map out your features on grid paper and record your calculations in your Math Journal."*

Show students the example park on the front of the In-Depth Inquiry Based Task card and give them three minutes to discuss initial ideas with a partner.

Then facilitate a group discussion about what needs to happen to make their dream designs and what sort of math work will need to be done. They might raise issues such as:

- What dimensions can we use for our park?
- Does our park have to be shaped like a rectangle or square? How about our park features?
- What scale should we use?
- Where should we write the dimensions of the features?
- How should we record our math thinking for finding the perimeters, areas, and lengths?

Day 1 (continued)

Encourage students to determine the answers to their questions. Discuss how to determine a reasonable scale. Note: Students should provide all their measurements in meters.

SAY *"Your design plans will be like looking down from an airplane, so you will not need to provide the height but will need to label the length and width of each feature. You must record your scale and keep track of the measurements and calculations in your Math Journal. You can use any of the perimeter and area of rectangles formulas learned in Unit 4 to show the perimeter and area of each feature."*

As a group, model a few possible features on a grid. Discuss ways to decide on a scale, options for labeling the features and their dimensions, and the best format for recording the calculations used to find the perimeters and areas of the features.

Give students time to plan their park features and the features' dimensions in their Math Journals.

Day 2 COMPLETE THE TASK

Gather the group and remind students of Day 1's work. Distribute Math Journals and copies of the Centimeter Grid Paper. Students might also need tape to put together more than one piece of grid paper.

Have students begin drawing the features of their dream parks based on the prior day's plans, encouraging them to work neatly and label features clearly. Remind students to record any math thinking in their journals.

Once the park plans have been completed and their sketches and calculations are recorded, discuss students' finished plans and ask students to share what they recorded in their journals. Possible discussion questions might include:

- I see that some of your parks are different sizes than others. How is that possible if you were to all have a park of 2,000 square meters?
- For shape features was it easiest to find the area and perimeters? Why?
- For what features was it hardest to find the area and perimeters? Why?
- How did knowing the formulas for finding the area and perimeters affect your choices in features and designs?
- How did designing your park on a grid help you measure and find the area and perimeters of your features?
- Did any of you have features for which you had to draw a line on a diagonal? Will you share how you calculated that perimeter and area?

SAY *"Close your eyes and think about how old you are in years. How can you convert that age into how old you are in months, weeks, days, and even hours? Write about it in your Math Journal."*

Grade 4, Unit 6 | Piece Offering

What is the objective?

Use knowledge of equivalent fractions and operations of fractions to solve a real-world problem.

Materials

- Math Journals (BLM page 55)
- Construction paper
- Glue
- Scissors

Time

40 minutes per day for two days

NOTE TO TEACHERS English language learners might be confused by the word play in the title of this task. Consider explaining to them the homophones piece and peace.

Day 1 INITIATE INQUIRY

Open by asking students if they have ever made homemade pizza.

SAY *"When I give the signal, turn and talk to a neighbor about what you need to make pizza and what your favorite toppings are."*

Give a signal to turn and talk, have students spend one minute discussing, and then gain their attention again to elicit background knowledge.

Announce that the class will be making paper pizzas during this two-day task.

SAY *"Picture yourself in a pizza predicament. Pretend that you and your three friends only have enough dough to make one pizza. Each of you wants different toppings. One friend wants twice as many slices as the others. All the slices need to be the same size."*

Give students a minute to turn and talk to their neighbors about the activity.

SAY *"Today, in groups of three or four, you will make a plan for a pizza that will solve the pizza predicament. Tomorrow, using construction paper, scissors, and glue, you will make the paper pizza. You will need to record your thinking in your Math Journal."*

Show students a sample construction paper pizza slice you created. Discuss common pizza toppings, writing them on the board for later use. Students may make rectangular or circular pizzas.

Give each student a copy of the Math Journal. Organize students into groups of two or three to engage and write initial attempts at solving the problem.

Day 1 (continued)

Then facilitate a group discussion about the math models they used in Unit 6 to solve fraction problems.

Fraction Bars

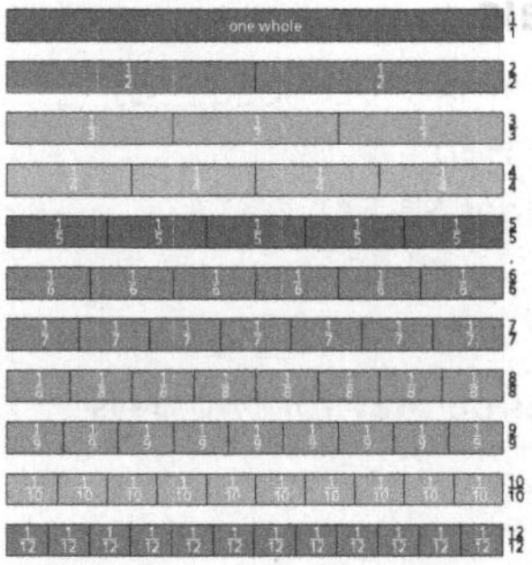

Fraction Models

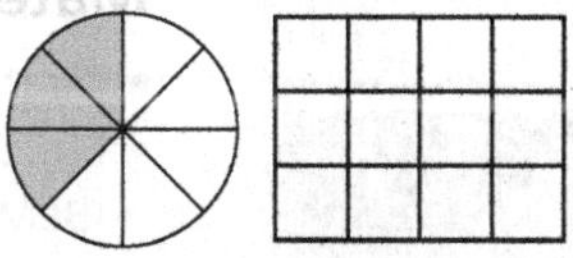

Ask three or four students to share their math thinking so far. As students speak, use math models to illustrate what they say if they do not already have math models to show.

SAY *"Next, you will work with your group again and add math models to your journals to best explain your thinking."*

Students then finish solving the problem and planning their pizzas. If students cannot figure out a solution that doesn't result in extra slices, nudge them to use the math models to visualize how many slices the pizza would need.

Day 2 COMPLETE THE TASK

Gather the class and remind the students of Day 1's discussion and main takeaways.

Ask students to quietly think for one minute about the pizza fractions problem, look back at the notes they took in their Math Journals, and imagine the toppings they will use for the four different types of pizza. After a minute has passed, ask them to turn and talk to a neighbor about their thoughts.

Give students time to make the paper pizzas and record their thinking in the journals. Act as a consultant, answering questions and suggesting math models when groups get stuck.

Once the pizzas are made, display them around the classroom and give students time to see their classmate's solutions. Gather the class together and ask a few students to share what they recorded in their journals. Possible discussion questions might include:

- How many slices of pizza do you and each of your friends get?
- How do you know each received the share of the pizza they wanted?
- What challenges did you encounter when you first tried to solve the problem?
- What would happen to the size of each slice if one more friend came to eat pizza?
- What other ways could you cut the pizza?
- What would happen to the fractions if you cut each slice in half? What equivalent fractions would that show?

SAY *"Imagine you are planning a pizza party for the teachers at school. How many pizzas will you need to order? What fraction of a pizza will each teacher eat? What fraction of all the pizzas needs to be vegetarian? Plan a teacher pizza party and write about it in your Math Journal."*

Grade 4, Unit 7 | Fraction Hop

What is the objective?
Use knowledge of fractions to solve a real-world problem.

Materials

- Math Journals (BLM page 55)
- Fraction Hop (BLM pages 53–54)

Time

30 minutes per day for two days

NOTE TO TEACHERS The lesson below is written as an indoor task, but the activity could be done outside, as the card photo suggests.

Day 1 INITIATE INQUIRY

Open by asking students if they have ever played board games or outdoor games such as hopscotch.

SAY *"When I give the signal, turn and talk to a neighbor about a time you played hopscotch and how the game is played. Or talk about your favorite board game."*

Give a signal to turn and talk, have students spend one minute discussing, and then gain their attention again to elicit background knowledge.

Announce that the group will be inventing a new twist on hopscotch or a new board game. The games must include fractions! Make up the rules and a game objective. Show students the paper hopscotch court and game path. Discuss how the games are normally played for students unfamiliar with them.

SAY *"Let's say you choose hopscotch. In each square, you could write a fraction. Players could choose fraction cards from a deck and hop their game markers to equivalent fractions on the course. For example, if you draw a card that says $\frac{12}{18}$, you would hop your marker to a hopscotch square that reads $\frac{2}{3}$."*

Give students time to brainstorm game ideas with a partner. Remind them to incorporate fractions into their game rules and instructions. Provide students with Math Journals so they can record their ideas.

NOTE If students can't think of an idea and want to use the example idea, you could provide the following additional instructions:

- The object of the game is to collect fraction cards by drawing a card, hopping to a

Day 1 (continued)

fraction on the court that is equivalent to the one on the card, and then hopping back to the start.

- If there is no fraction on the court equivalent to the one on the card you draw, then you must name an equivalent fraction. Correctly hopping to an equivalent fraction, or naming an equivalent fraction lets you keep that fraction card.
- If you hop to the wrong fraction, or cannot name an equivalent fraction, the card gets passed to the next player.
- The player with the most fraction cards, when time for playing is over or there are no more cards in the pile, wins!

Gain the group's attention and discuss the math models used in Unit 7 to compare fractions and to determine equivalent fractions, for example, fraction bars.

Ask three or four students to share how they plan to use fractions and equivalent fractions in their games. As the students speak, use math models to illustrate what they say if they do not already have math models to show.

SAY *"Tomorrow you will make your game board and game cards and play your game with other students."*

Day 2 COMPLETE THE TASK

Gather the group and remind students of Day 1's work. Have individuals or partners make their fraction games and game parts. Then have students teach each other how to play their games.

Once all students have had a chance to teach and play their games, discuss how it went. Questions might include:

- I see some of you played your games very rapidly. What's your secret for finding equivalent fractions so quickly?
- Which, if any, fractions were the most difficult to name equivalent fractions for?
- What was your favorite game to play? Which was the most challenging?
- If you had to design a game all over again, what would you change? Why?

SAY *"Think about other playground games. How could you change your favorite game to involve fractions? Could you dribble out equivalent factions with a basketball? Could you call out fractions during a game of Red Rover? Write about it in your Math Journal."*

Grade 4, Unit 8 | Wrangle Some Angles

What is the objective?
Use knowledge of angles to identify and measure them in real-world settings.

Materials

- Math Journals (BLM page 55)
- White paper
- Colored pencils
- Highlighter markers
- Clipboards
- Protractors
- Optional: digital cameras, iPads, and computer printer
- Unwanted magazines and newspapers

Time

40 minutes per day for two days

Day 1 INITIATE INQUIRY

NOTE This lesson is written as a drawing activity that students complete inside and outside their school. If it's not possible to take students to the playground, pictures from unwanted magazines and newspapers could be used instead. Also, instead of drawing pictures of objects, students could photograph items using mobile devices and find the angles on printed photos to complete the task.

Open by asking students if they have ever noticed how angles can be found everywhere.

SAY *"When I give the signal, turn and talk to a neighbor about the different angles they see around the classroom."*

Give a signal to turn and talk, have students spend one minute spotting angles, and then gain their attention again to elicit background knowledge.

Announce that today the group will predict the angles they will find at their school and tomorrow they will hunt for angles and record them in more accurate drawings.

SAY *"Think about the angles you will find on the playground, in the library, in the cafeteria, and in other places. Your first step is to draw quick sketches of objects you know are in those places and use a protractor to predict the sizes of the angles they form."*

Tell students that tomorrow they will visit those locations, draw a more accurate picture next to their prediction sketch, and see how far off their angle estimates were.

Provide students with white paper, colored pencils, protractors, highlighter markers, and Math Journals. In their journals, they should make a table of their angles. In one column, they will write the measurements of their angle predictions. In the second column, they will list the more accurate measurements from Day 2.

Math Activity Center

In-Depth Inquiry-Based Tasks *(continued)*

Day 1 (continued)

Give students time to sketch and predict angles. They should trace over the angles they are measuring with the highlighter markers.

Act as a consultant during the highlighting and measuring activity, answering questions and making sure students use the protractor correctly.

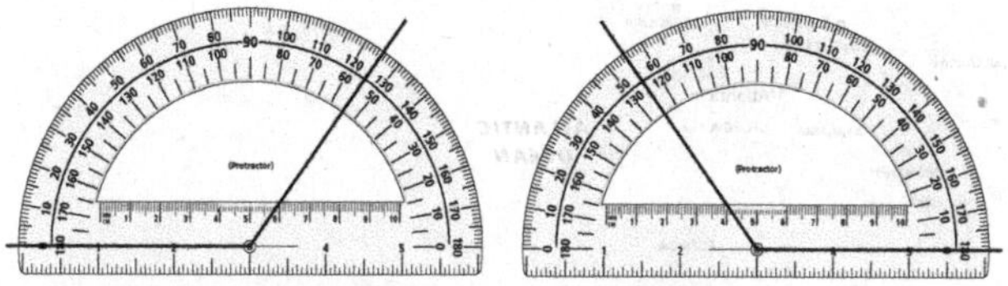

Use the outer scale. Use the inner scale.

Day 2 COMPLETE THE TASK

Gather the group and remind students of Day 1's work. Give students their journals and supplies from the day before.

SAY *"Today you will go to the places around the school where you drew angles from memory yesterday. On white paper, you will make careful drawings of the objects you see and the angles they form. When you get back to the classroom, you will use your protractor to measure the new angles."*

Once the angles have been drawn, highlighted, and measured, have groups organize their angle photos by number of degrees, least to greatest. Display the angles around the classroom or on a bulletin board.

Discuss the activity with the group:

- What was the smallest angle you found? Greatest angle? Where did you find it?
- Did anyone find a right angle? Where? Are all the right angles the same size?
- Were most of the angles you found acute or obtuse?
- Did anyone find a straight angle? Explain.
- Did any of you find more than one angle in a single object?
- Did your perspective (head-on versus side-view) affect the measure of the angle?

SAY *"Now that you found and measured real-world angles, make your own angles. Choose an object from one of your photos, and draw that object in your Math Journal. Then draw that object two more times, once by making that angle smaller, and then again by making that angle larger than the original angle in your picture. Measure and label those three angles. How does changing the angle affect the shape of the object?"*

Name Date

Concert Tour Info Sheet

Anchorage:
Stadium capacity: 66,387
Stadium cost: $12,450
Hotel: $99 per night
Food: $38 per day

Denver:
Stadium capacity: 112,273
Stadium cost: $31,234
Hotel: $142 per night
Food: $53 per day

Austin:
Stadium capacity: 135,119
Stadium cost: $29,500
Hotel: $164 per night
Food: $49 per day

Seattle:
Stadium capacity: 122,392
Stadium cost: $32,100
Hotel: $201 per night
Food: $52 per day

Phoenix:
Stadium capacity: 83,400
Stadium cost: $14,045
Hotel: $110 per night
Food: $42 per day

Detroit:
Stadium capacity: 78,297
Stadium cost: $20,000
Hotel: $111 per night
Food: $37 per day

Philadelphia:
Stadium capacity: 115,176
Stadium cost: $36,050
Hotel: $175 per night
Food: $48 per day

Nashville:
Stadium capacity: 102,143
Stadium cost: $30,754
Hotel: $175 per night
Food: $45 per day

New York:
Stadium capacity: 212,864
Stadium cost: $50,021
Hotel: $270 per night
Food: $100 per day

Atlanta:
Stadium capacity: 138,258
Stadium cost: $15,678
Hotel: $86 per night
Food: $40 per day

Minneapolis:
Stadium capacity: 76,665
Stadium cost: $19,900
Hotel: $95 per night
Food: $32 per day

Honolulu:
Stadium capacity: 128,490
Stadium cost: $43,128
Hotel: $205 per night
Food: $78 per day

Name ______________________ Date ______________________

Zoo Trip Planner

Budget = $ ________

Cost per bus is $150. Each bus holds 50 people.

Lunch costs $8 per child and $10 per adult.

Total cost for buses:	Total cost for lunches:

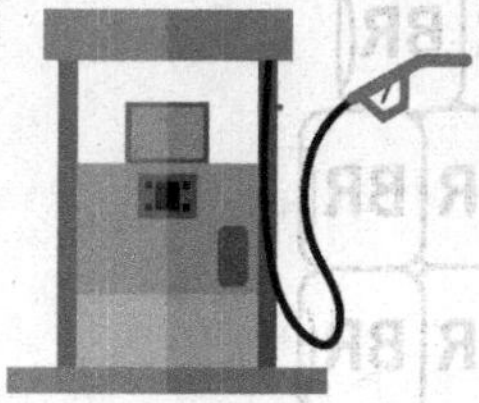

Cost of gas is $2.50 per gallon. The bus can travel 5 miles per gallon.

Zoo tickets cost $12 per student and $14 per adult.

Total cost for gas:	Total cost for tickets:

Price of Trip = $ ________

The Great Bead Divide

Use the information below to find out how many of each type of beaded animal you can make.

You have:

48 red beads (R)

22 pink beads (P)

112 yellow beads (Y)

170 green beads (G)

128 blue beads (BU)

121 white beads (W)

84 tan beads (T)

441 brown beads (BR)

119 black beads (BK)

160 purple beads (PR)

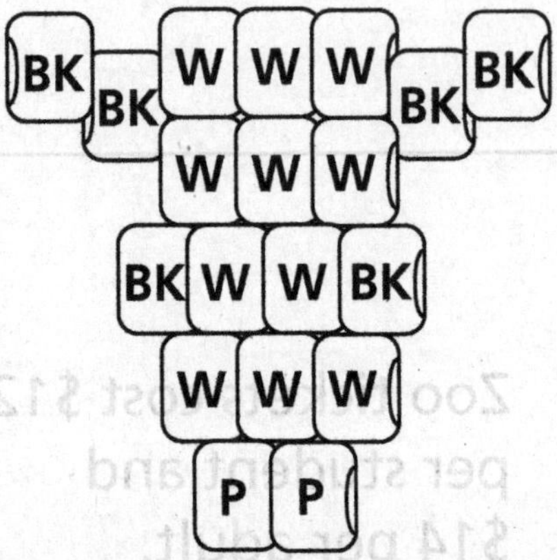

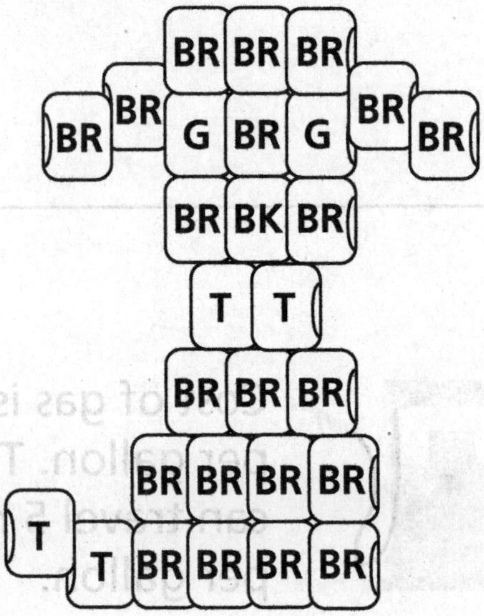

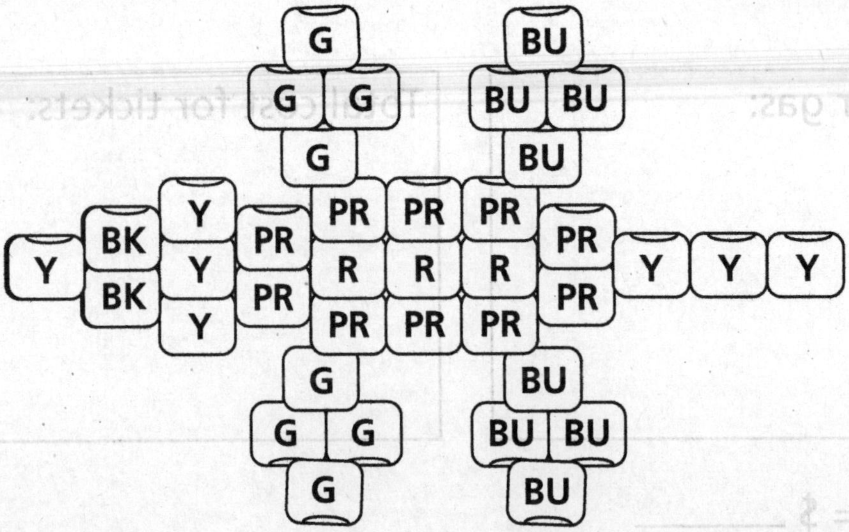

Name Date

Buy the Book Info Sheet

Use the information below to plan Longfellow School's book fair.

Remember to order enough books so that each child has a choice of at least five. Order books that children from all grade levels will like.

It costs the school $2 to buy each book, and then the school sells each book for $6. How much will it cost the school to buy the books? How much does the school earn off the sale of each book?

Grade level	Number of students
Kindergarten	43
First grade	65
Second grade	59
Third grade	74
Fourth grade	56
Fifth grade	78

If every student buys three books, how much money will the school earn?

The school needs $6,000 to buy new musical instruments. Does selling three books per student earn enough money? How many more books would each student need to buy to reach that goal?

Draw a map of the book-sale space. If you have 9 tables, how many books will need to go on each table?

Name **Date**

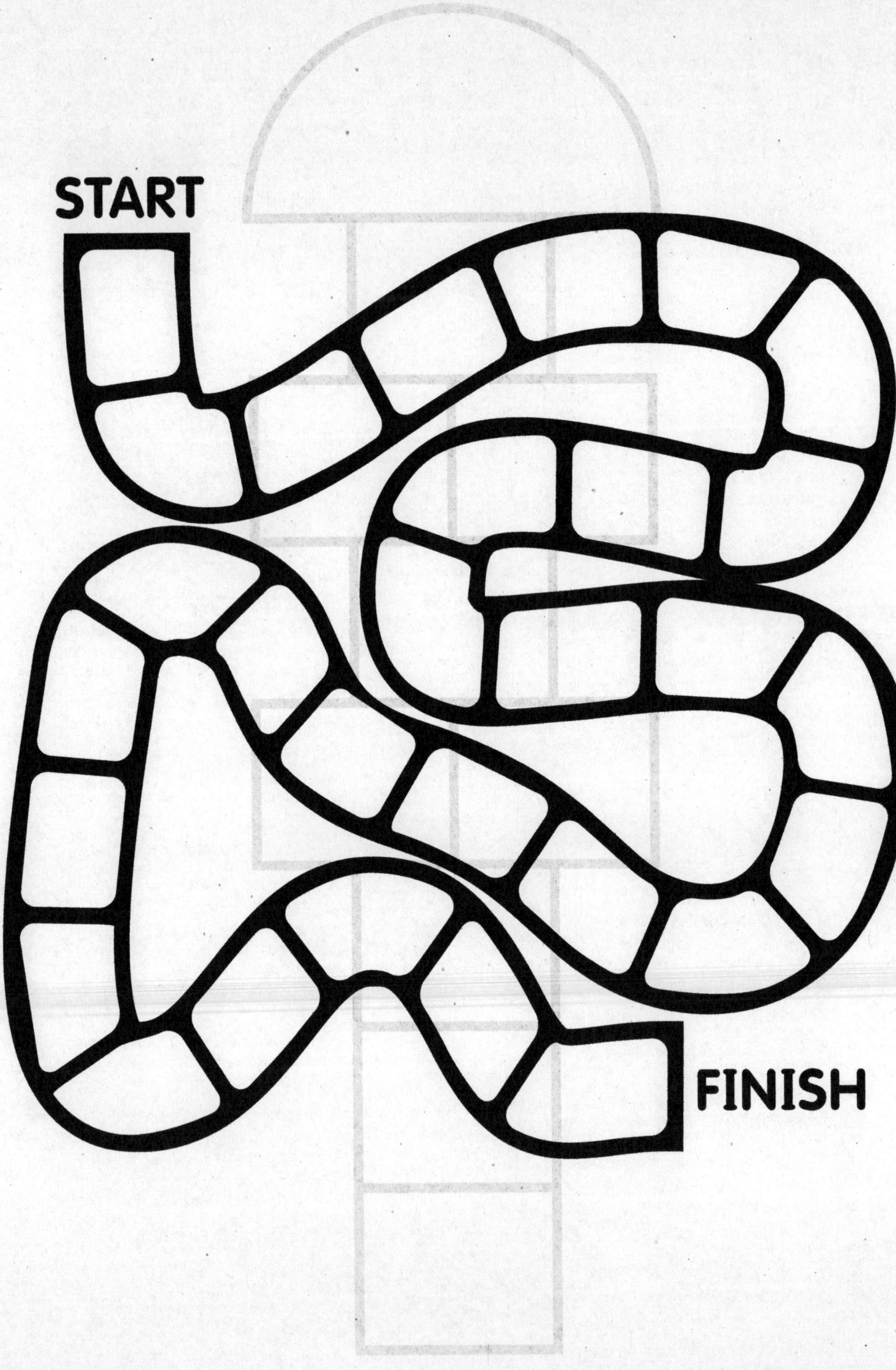
START
FINISH

Math Journal

Name ______________________ Date ____________

Unit Number ____________

Notes:

Math Readers

Math Readers bring mathematical concepts to life through fiction and nonfiction stories that engage the learner. In each unit's unique *Math Reader*, students will connect math to everyday life, delightful characters, and historic events. Interactive questions throughout the readers focus on key concepts and support student comprehension. A responding page, at the end of each reader, includes questions and an activity for additional practice after the reading.

Teacher Notes for *Math Readers* include:

- key skills and concepts
- story summary
- essential vocabulary words
- responding/activity answers

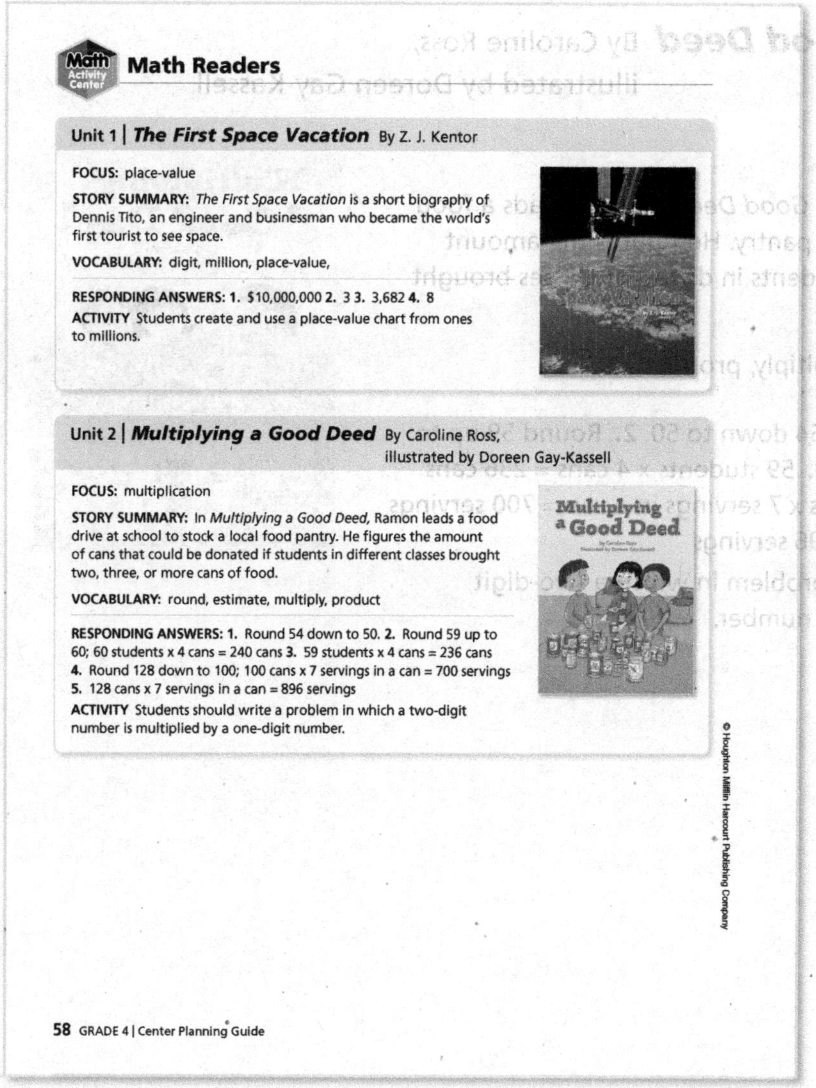

Math Readers

Unit 1 | *The First Space Vacation* By Z. J. Kentor

FOCUS: place-value

STORY SUMMARY: *The First Space Vacation* is a short biography of Dennis Tito, an engineer and businessman who became the world's first tourist to see space.

VOCABULARY: digit, million, place-value,

RESPONDING ANSWERS: 1. $10,000,000 **2.** 3 **3.** 3,682 **4.** 8

ACTIVITY Students create and use a place-value chart from ones to millions.

Unit 2 | *Multiplying a Good Deed* By Caroline Ross, illustrated by Doreen Gay-Kassell

FOCUS: multiplication

STORY SUMMARY: In *Multiplying a Good Deed*, Ramon leads a food drive at school to stock a local food pantry. He figures the amount of cans that could be donated if students in different classes brought two, three, or more cans of food.

VOCABULARY: round, estimate, multiply, product

RESPONDING ANSWERS: 1. Round 54 down to 50. **2.** Round 59 up to 60; 60 students x 4 cans = 240 cans **3.** 59 students x 4 cans = 236 cans **4.** Round 128 down to 100; 100 cans x 7 servings in a can = 700 servings **5.** 128 cans x 7 servings in a can = 896 servings

ACTIVITY Students should write a problem in which a two-digit number is multiplied by a one-digit number.

Math Readers

Unit 1 | *The First Space Vacation* By Z. J. Kentor

FOCUS: place-value

STORY SUMMARY: *The First Space Vacation* is a short biography of Dennis Tito, an engineer and businessman who became the world's first tourist to see space.

VOCABULARY: digit, million, place-value,

RESPONDING ANSWERS: 1. $10,000,000 **2.** 3 **3.** 3,682 **4.** 8

ACTIVITY Students create and use a place-value chart from ones to millions.

Unit 2 | *Multiplying a Good Deed* By Caroline Ross, illustrated by Doreen Gay-Kassell

FOCUS: multiplication

STORY SUMMARY: In *Multiplying a Good Deed,* Ramon leads a food drive at school to stock a local food pantry. He figures the amount of cans that could be donated if students in different classes brought two, three, or more cans of food.

VOCABULARY: round, estimate, multiply, product

RESPONDING ANSWERS: 1. Round 54 down to 50. **2.** Round 59 up to 60; 60 students x 4 cans = 240 cans **3.** 59 students x 4 cans = 236 cans **4.** Round 128 down to 100; 100 cans x 7 servings in a can = 700 servings **5.** 128 cans x 7 servings in a can = 896 servings

ACTIVITY Students should write a problem in which a two-digit number is multiplied by a one-digit number.

Unit 3 | ***The Division Champs*** By G. Ramirez, illustrated by Michelle Dorenkamp

FOCUS: division

STORY SUMMARY: The Midway Lions are practicing for the School Quiz and decide to practice division with remainders. When the day of the competition arrives, they are prepared for all the questions, even the game-winning division question.

VOCABULARY: division with remainders, dividend, divisor, quotient, remainder

---

RESPONDING ANSWERS: 1. 3 **2.** 11 **3.** Each team member got 9 CDs; there were 2 left over.

ACTIVITY 21 R1; 9 R2; 4 R1; 31 R1

Unit 4 | ***Eratosthenes and His Sieve*** By A. I. Freeman, illustrated by Wendy Born Hollander

FOCUS: prime and composite numbers

STORY SUMMARY: *Eratosthenes and His Sieve* is a short biography of Eratosthenes and his amazing contributions to math, including his sieve for identifying prime numbers.

VOCABULARY: prime number, factor, composite number

RESPONDING ANSWERS: 1. You get a composite number because 2 is a factor of it, as well as 1 and the number itself. **2.** You get a composite number because 5 is a factor of it, as well as 1 and the number itself. **3.** You get a composite number because 4 is a factor of it, as well as 1, 2, and the number itself. **4.** 81 is a composite number: $9 \times 9 = 81$. **5.** 101

ACTIVITY 1. They are all odd numbers except for 2. Each prime number has only itself and 1 as factors. **2.** They are not in any order. **3.** It is an even number and has more than two factors: 1, 2, 3, 6, 9, and 18. **4.** It has only itself and 1 as factors. **5.** Composite numbers have more than two factors.

Unit 5 | *Measuring the Mississippi* By Marilyn Eden

FOCUS: measuring length

STORY SUMMARY: *Measuring the Mississippi* is written as a travel brochure for a paddleboat trip down the Mississippi River. Different lengths of measurement are used to discuss various aspects of the trip.

VOCABULARY: foot, yard, mile, centimeter, meter, kilometer, system of measurement, customary, metric

---

RESPONDING ANSWERS: 1. United States—customary, rest of the world—metric **2.** kilometers

ACTIVITY Students should use a ruler to measure the animals.

Unit 6 | *A Melody in Fractions* By Ellen Philips, illustrated by Nancy Lane

FOCUS: equivalent fractions

STORY SUMMARY: In *A Melody in Fractions,* Sachiko is learning to read music and play the piano. Her teacher uses fractions to help her understand whole notes, half notes, quarter notes, eighth notes, and sixteenth notes.

VOCABULARY: fractions, equivalent fractions, fraction strips, denominator, numerator

---

RESPONDING ANSWERS: 1. 1/8 **2.** 8 **3.** B **4.** a shorter time **5.** 4

ACTIVITY Students should make fraction strips as indicated.

Math Readers *(continued)*

Unit 7 | ***Sleeping Half the Day Away*** By Carolina Falls

FOCUS: fractions

STORY SUMMARY: Some animals sleep only a few hours per day while others sleep more than half the day. In *Sleeping Half the Day Away,* students learn what fraction of the day different kinds of animals sleep.

VOCABULARY: fraction, equivalent fractions, denominator, numerator, order, compare

RESPONDING ANSWERS: 1. 7/12 **2.** 1/4; 5/12; 2/3 **3.** 4/6; 16/24; 2/3

ACTIVITY Elephant 1/6; Gerbil 13/24; Squirrel 5/8; Koala 5/6

Unit 8 | ***Skateboarding Takes Shape*** By Loretta West

FOCUS: shapes and angles

STORY SUMMARY: *Skateboarding Takes Shape* discusses the shapes often used in a skatepark and the angles of some skateboarding tricks and jumps.

VOCABULARY: angle, acute angle, intersecting lines, parallel lines, perpendicular lines, triangle, rectangle, polygon

RESPONDING ANSWERS: 1. B **2.** A **3.** C

ACTIVITY Students should draw a skatepark or funbox with at least one triangle, one rectangle, and a set of parallel lines.